Bioaccumulation of Xenobiotic Compounds

Des W. Connell, Ph.D.
Division of Australian Environmental Studies
Griffith University
Brisbane, Australia

CRC Press
Taylor & Francis Group
Boca Raton London New York

CRC Press is an imprint of the
Taylor & Francis Group, an **informa** business

First published 1990 by CRC Press
Taylor & Francis Group
6000 Broken Sound Parkway NW, Suite 300
Boca Raton, FL 33487-2742

Reissued 2018 by CRC Press

© 1990 by CRC Press, Inc.
CRC Press is an imprint of Taylor & Francis Group, an Informa business

No claim to original U.S. Government works

Library of Congress Cataloging-in-Publication Data

Connell, Des W., 1938-
    Bioaccumulation of xenobiotic compounds / Des W. Connell.
        p. cm.
    Bibliography: p.
    Includes index.
    ISBN 0-8493-4810-2
    1. Xenobiotics--Bioaccumulation. I. Title.
QH4545.X44C66 1989
574.2'4--dc19   89-954

A Library of Congress record exists under LC control number: 89000954

Publisher's Note
The publisher has gone to great lengths to ensure the quality of this reprint but points out that some imperfections in the original copies may be apparent.

Disclaimer
The publisher has made every effort to trace copyright holders and welcomes correspondence from those they have been unable to contact.

ISBN 13: 978-1-315-89102-6 (hbk)
ISBN 13: 978-1-351-07012-6 (ebk)

Visit the Taylor & Francis Web site at http://www.taylorandfrancis.com and the
CRC Press Web site at http://www.crcpress.com

PREFACE

Widespread public concern regarding xenobiotic chemicals in the environment probably began in 1962 with the publication of Rachel Carson's book *Silent Spring*. A number of undesirable effects were attributed to agricultural pesticides, particularly DDT, which were previously unsuspected by all but a few specialized scientists. Since then a host of problems associated with the use of chemicals have emerged: human health problems related to herbicide use, declining populations of some eagle species related to DDT, PCB contamination of seafoods, and so on. Today the general public is highly conscious of chemical contamination of the environment, so much so that a new word, "chemophobia", has been coined to describe this phenomenon in its most extreme form.

Governments throughout the world have responded to this concern by instituting programs to monitor the distribution and effects of existing chemicals, and to evaluate environmental effects of any new chemicals. Action has also occurred at an international level with the Organization for Economic Cooperation and Development (OECD), which has coordinated the development of an internationally accepted chemical evaluation program.

Central to these concerns is the potential of some xenobiotic chemicals to transfer from the external environment to biota with an increase in concentration. The increases in concentration in some cases, for example, DDT, have been found to be substantial. Thus a chemical, at concentrations in water, soil, or air which are considered to be harmless, may appear at dangerous concentrations in biota. The observed sequential increases in the concentrations of some chemicals in food chains have become part of the general knowledge of most people in the community.

Parallel with this increase in public and governmental interest, there has been an extensive research program on the bioaccumulation of xenobiotic chemicals. Considerable advances have been made in recent years in the understanding of bioaccumulation, and a major body of information is now available. However, there are many aspects which are not well understood and remain to be resolved by future research. The first objective of this book was to collate the current information and present an evaluation of the state of knowledge on bioaccumulation in a concise form, rather than scattered throughout an assortment of journals, books, conference proceedings, and so on. A second objective was to evaluate relationships and knowledge of mechanisms by bringing together approaches and data from several sources which individually may not have had a clear meaning. The final objective was the development of a knowledge of where there are critical gaps in our understanding.

In preparing this book my colleague, Darryl Hawker, assisted in many ways, but particularly by kindly agreeing to write two chapters. My wife Patricia was of considerable help in collecting many papers from various libraries. The rather difficult manuscript was patiently and efficiently typed by Cheryl Sheehan and Carolyn Plant; Aubrey Chandica skillfully prepared the diagrams, and Scott Bryne assisted in obtaining permission to use material from publishers. To all of these people I am extremely grateful for their willing and helpful assistance, without which the book could not have been prepared.

Des W. Connell
Brisbane

THE AUTHOR

Des W. Connell is a Reader in environmental chemistry in the Division of Australian Environmental Studies at Griffith University and Director of the Government Chemical Laboratory in Brisbane, Australia. His broad interests are in environmental chemistry, ecotoxicology, and water pollution, and he has published several books on these topics. His research interests are concerned with the behavior of xenobiotic chemicals in the environment. In particular, he has published papers on the prediction of bioconcentration of persistent lipophilic chemicals by aquatic biota, bioconcentration kinetics, and the physicochemical characteristics of various lipophilic compounds. Before joining Griffith University, he was Marine Studies coordinator of the Westernport Bay Environmental Study and Director of the Gippsland Lakes Environmental Study with the Victorian Ministry for Conservation. Currently, he is a member of the Australian Environment Council's Chemical Review Subcommittee and Chairperson of the Environmental Chemistry Division of the Royal Australian Chemical Institute. Dr. Connell received B.Sc., M.Sc., and Ph.D. degrees in Organic Chemistry from the University of Queensland.

CONTRIBUTOR

Darryl W. Hawker is a Teaching Fellow in environmental chemistry in the Division of Australian Environmental Studies at Griffith University. His research interests are in the physicochemical properties of chemicals, particularly organometallic compounds, and how these properties influence environmental behavior. Dr. Hawker received his B.Sc. (Hons) and Ph.D. degrees in Organic Chemistry from the University of Queensland.

TABLE OF CONTENTS

Chapter 1

INTRODUCTION

Des W. Connell

TABLE OF CONTENTS

I. XENOBIOTIC CHEMICALS AND BIOACCUMULATION

Xenobiotic chemicals are chemicals foreign to life, which are usually derived synthetically or from an abiotic process. The term "xenobiotic" is a combination of the Greek words "xenos", meaning strange or foreign, and "bios", meaning life. Thus xenobiotic chemicals are pollutants in the biosphere although not all pollutants are xenobiotic chemicals. The synthetic xenobiotic chemicals are often of enormous value to human society, and are usually the majority of the chemicals in such important groups of substances as pharmaceuticals, petrochemicals, pesticides, and plastics. Schmidt-Bleek and Haberland[1] in 1980 reported that there were 40,000 chemicals commercially available, and 2,000 were placed on the market in addition each year.

Since the very beginning of the chemical industry there has been interest in producing more efficacious products. This has led to continuing research into the prediction of the likely properties of a chemical prior to its use. Concurrently, research has been in progress which will give a better understanding of the mode of action of chemicals. One of the most important properties of a chemical, in situations involving a biological effect or application, is how well it is absorbed or bioaccumulated. Bioaccumulation usually means the accumulation of a chemical in an organism to a higher concentration than is present in an external source.

II. HISTORICAL DEVELOPMENT OF KNOWLEDGE ON ABSORPTION, BIOACCUMULATION AND RELATED EFFECTS OF XENOBIOTIC COMPOUNDS

The basis of our understanding of bioaccumulation was developed by scientists working in the 1870's who discovered some of the principal phenomena governing the behavior of chemicals in gases, water, and other phases. Some of these principles are now used in contemporary investigations of the behavior of chemicals in the environment and in organisms. For example, Berthelot and Jungfleisch[2] in 1872, and Nernst[3] in 1891, discovered that a constant partition coefficient, or "distribution ratio", controlled the partitioning of a single pure chemical between two different phases. This knowledge marks an important step forward, since once the partition coefficient is known for a compound, its distribution between the same phases in other situations can then be calculated. The bioaccumulation of a chemical by an organism can be seen, in many situations, as a partition process.

During this early time period, interest was not only developing in aspects of chemical behavior, but also in the behavior of drugs, poisons, and narcotics in organisms. Thus, during this pioneering period, others were building on the basic physicochemical knowledge available, and utilizing this in attempting to elucidate the factors governing the physiological properties of chemical compounds. Probably the earliest among these were Crum-Brown and Fraser,[4,5] who in 1868-1869 proposed that the physiological action of a compound was a function of its chemical constitution. Later in 1893, this found practical expression in the work of Richet,[6] who found that the toxic effects of certain ethers, alcohols, aldehydes, and ketones were inversely related to their solubility in water.

A few years later, interest was first kindled in the partition behavior of nonelectrolyte chemicals in relation to their physiological effects on organisms. The English born and Swiss-trained Charles Ernest Overton commenced his research program into the relationship of physicochemical properties of compounds to their narcotic effect on organisms virtually single-handedly. Also about this time, Hans Meyer and his collaborator Fritz Baum were active in the same field. A series of papers[7,8,9] were published by these gifted researchers culminating in 1901 with the publication of Overton's book, *Studien Uber Die Narkose*[10] (*Studies of Narcosis*). This book is a landmark in studies of how physicochemical properties of chemicals influence their physiological effects on organisms. It sets out many of the basic principles which have been continually refined and expanded up to the present day.

One of the major discoveries of these researchers was that the narcotic action of a nonelectrolyte is correlated with the compound's lipoid substance-to-water partition coefficient. Overton's studies were carried out with 130 compounds including alcohols, hydrocarbons, nitriles, nitroparaffins, aldehydes, and ketones, and with a variety of test animals including tadpoles, fish, crustacea, and Daphnia.

In 1904, Traube[11] extended the physicochemical properties of interest from the partition coefficient to surface tension when he established a linear relationship between surface tension and activity for a series of narcotics. Aspects of structural organic chemistry were introduced by Fuhner,[12,13] who noted a possible relationship between narcotic activity and the number of carbon atoms in a compound.

A considerable volume of data was accumulated in the following years, confirming the general accuracy of the previously developed relationships. However, during the 1950s and 1960s further significant refinements and developments occurred. Hansch and Leo,[14] together with a variety of co-workers, suggested the use of the octanol to water partition coefficient (K_{ow}) as the most suitable partitioning phase pair for studying the relationship between partition coefficient and the biological properties of compounds. They systematized much of the information available into quantitative structure activity relationships (QSAR).

Initially, this group developed the following equation for toxicity:

$$\log (1/C) = k_1 \pi + k_2 \sigma + k_3$$

where C is the molar concentration of a compound producing a standard response in a constant interval; k_1, k_2, and k_3 are empirical constants; π is $\log P_X - \log P_H$, where P_X is the octanol-to-water partition coefficient of a derivative and P_H is that of a parent molecule; and σ is the Hammett constant.

When k_2 is equal to zero, this equation is equivalent to the Meyer-Overton relationship, which predicts a linear relationship between $\log(1/C)$ and the octanol to water partition coefficient. When k_2 has an empirical value, the Hammett constant is included in the relationship and provides a much better correlation for many compounds with toxicity than the Meyer-Overton relationship mentioned above. It is noteworthy that the relationship between the Hammett constants alone and toxicity is not particularly strong. Later, this linear relationship was developed by Hansch[15] into a parabolic form expressed by the following equation:

$$\log (1/C) = -k_1 (\log K_{ow})^2 + k_2 (\log K_{ow}) + k_3 \sigma + k_4$$

where k_{ow} is the octanol to water partition coefficient and k_1, k_2, k_3 and k_4 empirically developed constants. This relationship has been extensively and successfully used to explain the toxicity of a variety of compounds.

Up to the period of these developments, studies of the uptake and behavior of xenobiotic chemicals in organisms was principally concerned with relatively short-term effects, such as toxicity and narcosis. The applications were mainly concerned with pharmaceuticals and their effectiveness in relationship to human health. In the 1960s widespread concern developed regarding the use of persistent chemicals in the environment. Much of this related to the chlorinated hydrocarbons, particularly DDT. These substances tend to have little effect in the short term, but trace residues bioaccumulated within organisms over long periods have had adverse effects, in many cases. As a result of this, there was a developing interest in the presence of persistent chemical residues in organisms, and the routes and mechanisms of entry of these substances. Accumulation through steps in the ecological food chain, with progressive increases in concentration, was initially proposed as the mechanism of entry for many organisms. With terrestrial ecosystems the presence of contaminants in food must be the principle source of residues in organisms. However, in aquatic ecosystems water to organism partitioning has been found to be a significant influence on the presence of residues in organisms within the system. In principle, it could be expected that the biological response of bioaccumulation with environmental xenobiotic compounds would be related to the factors generating the biological response observed with pharmaceutical xenobiotic compounds such as drugs, narcotics, and so on. Thus, it could be expected that the bioaccumulation of environmental xenobiotics would be related in some way to the octanol-to-water partition coefficient of the compounds involved.

The general characteristics of the interaction of xenobiotic chemicals with organisms can be seen in simplified terms as expressed in Figure 1. The basic characteristics of a molecule, such as structure, number of carbon atoms, and so on, govern its physicochemical properties, such as chemical reactivity, water solubilities, and so on, and as a result of these the substance can be absorbed by an organism and possibly bioaccumulate. The amount of the chemical which enters the organism, together with its basic molecular characteristics and physicochemical properties, can all be expected to have an influence on the resultant biological activity.

III. IMPORTANCE OF BIOACCUMULATION IN ENVIRONMENTAL MANAGEMENT

Prior to the 1950s, the xenobiotic chemicals of principle interest were pharmaceutical chemicals, but since that time there has been increasing interest in the environmental xenobiotic chemicals. The bioaccumulation of a number of persistent chemicals discharged into the environment has had detrimental effects on organisms within natural systems and also on human health. Initially, concern was focused on agricultural chemicals, particularly the chlorinated hydrocarbon insecticides. These have been distributed over extensive areas of the environment, and with DDT and DDE, effects such as eggshell thinning in birds containing these residues have resulted in a lack of reproductive success. In more recent times, the use of the polychlorinated biphenyls (PCBs) in industry has been found to result in residues in some human populations and in natural ecosystems. A variety of deleterious effects have been ascribed to this situation. In the last few years, other chemicals, such as the polyaromatic hydrocarbons (PAH), dioxins, and dibenzofurans have been similarly observed to form residues with resultant adverse effects. The sources and environmental entry pathways are often complicated, and are not fully delineated in many cases. Thus, there has been general agreement that there should be an intensified investigation of all chemicals with potential adverse environmental effects. This has found expression in the initiation of schemes for the registration of chemicals before use. The U.S. Toxic Substances Control Act of 1976, and the OECD scheme for the hazard assessment of chemicals[16] are examples of these schemes. Important aspects of hazard evaluation are the bioaccumulation capacity of a chemical, and the prediction of this property from physicochemical properties measured in the laboratory.

It is interesting to note that development of concern regarding the environmental occurrence of xenobiotic chemicals is related to the development of scientific methods for chemical analysis. In 1952 James and Martin[17] invented the gas liquid chromatography technique which has principle application to the analysis of organic compounds. Later the technique was extended and improved such that trace amounts of xenobiotic organic chemicals could be quantified in environmental samples. The flame ionization detector extended the sensitivity of the method so that levels of a few parts per million for many organic compounds could be detected in samples. In addition, with the chlorinated hydrocarbons,

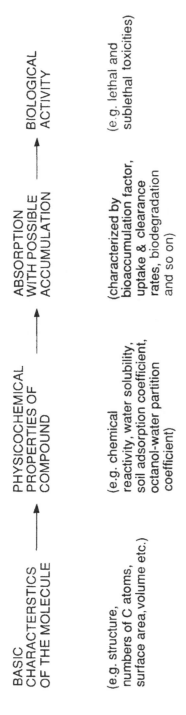

BASIC
CHARACTERSTICS
OF THE MOLECULE

(e.g. structure,
numbers of C atoms,
surface area, volume etc.)

PHYSICOCHEMICAL
PROPERTIES OF
COMPOUND

(e.g. chemical
reactivity, water solubility,
soil adsorption coefficient,
octanol-water partition
coefficient)

ABSORPTION
WITH POSSIBLE
ACCUMULATION

(characterized by
bioaccumulation factor,
uptake & clearance
rates, biodegradation
and so on)

BIOLOGICAL
ACTIVITY

(e.g. lethal and
sublethal toxicities)

FIGURE 1. General relationships between the various properties of a molecule and compound.

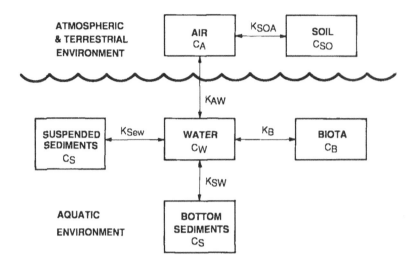

FIGURE 2.   A diagrammatic representation of the distribution of a chemical between phases in the environment, with the values indicating concentrations in a phase and the K values the partition coefficients at equilibrium.

the development of the electron capture detector further extended the sensitivity of the technique. In 1967, only 15 years after the first report on gas liquid chromatography, a considerable body of information was available on the concentrations of xenobiotic organic compounds in biota, as indicated by Woodwell.[18] At this time the explanation of the behavior of these compounds in the environment was based principally on the nature and relationship of these concentrations. For example, food chain transfer, as explained by Woodwell,[18] was believed to be the major mechanism operating. With this mechanism, food organisms are consumed by a predator containing food substances, such as carbohydrate, protein and fat, together with trace xenobiotic contaminants. In digestion and consumption the food substances are readily degraded and removed, leaving the resistant contaminant chemicals to be accumulated by the predator. Further repetitions of this process in a food chain could be expected to lead to progressive increases in concentration of a contaminant chemical, and thus the development of a relationship between the concentration of the contaminant and the position of the organism in the food chain.

More recently, the environment has been considered to be made up of a number of different phases, such as water, air, soil, and biota. When a chemical is discharged it distributes among these different phases according to the properties of the chemical and the phases,[19] as shown in Figure 2. At equilibrium the concentrations in each pair of phases are characterized by the partition coefficient K, which is the ratio of the concentration in one phase to the concentration in the other, and is constant. The water-to-biota partition coefficient is of particular importance because we are usually concerned with adverse effects on biota. The other aspects are principally of significance in the influence they have on this process. The water-to-biota partition coefficient K_B is often referred to as the Bioconcentration Factor, and can have values up to and above a million. Thus, aquatic biota may bioaccumulate significant concentrations of some substances from very low concentrations present within the water mass. In fact, in some cases the occurrence of contaminants in the water mass yielding significant biotic residues is so low that it cannot be detected by conventional analytical techniques.

The diagrammetric representation of the interaction of various properties and effects shown in Figure 1 implies a wider significance for bioaccumulation than simply the process which controls the presence of residues in organisms. The related processes of absorption and possible bioaccumulation precede the resultant biological activity, and are the processes controlling the actual amount of a chemical which enters an organism. The amount of a chemical in an organism, together with its other properties, give the final biological effect. In fact, in some cases with nonspecific toxicity, the amount of a substance is almost the sole control of the biological effect. Thus, a knowledge of the various factors affecting bioaccumulation should lead to an insight into factors affecting toxicity and other biological effects.

REFERENCES

1. **Schmidt-Bleek, F. and Haberland, W.**, The yardstick concept for the hazard evaluation of substances, *Ecotoxicol. Environ. Safety*, 4, 455, 1980.
2. **Berthelot, M. and Jungfleisch, E.**, Sur les lois qui president ou partage d'un corps entre deux dissolvants (experiences), *Ann. Chim.*, 26, 396, 1872.
3. **Nernst, W.**, Vertherlungeines Stoffes zwischen zwei Losungsnitteln und zwischer Losungsmittel und Dampfraum, *Z. Phys. Chem.*, 8, 110, 1891.
4. **Crum-Brown, A. and Fraser, T. R.**, On the connection between chemical constitution and physiological action, Part 1, *Trans. R. S. Edinburgh*, 5, 151, 1868-1869.
5. **Crum-Brown, A. and Fraser, T. R.**, On the connection between chemical constitution and physiological action, Part 2, *Trans. R. S. Edinburgh*, 25, 693, 1868-1869.
6. **Richet, M. C.**, Note sur le rapport entre la toxiatie et les properties physiques des corps, *Soc. Biol. Mem.*, 45, 775, 1893.
7. **Overton, E.**, Ueber die osmotischen eigen schaften der Zelle in ihrer dedeutung fur die Toxicologie und Pharmakologie., *Vierteljahrsschr. Narturforsch. Ges. Zuerich*, 41, 383, 1896.
8. **Meyer, H.**, Zur theorie der Alkoholnarkose. Erste mittheilung, welche eigenschaft der anasthetica bedingt ihre narkotiste wirkung, *Arch. Exp. Pathol. Pharmakol.*, 42, 109, 1899.
9. **Baum, F.**, Zur Theorie der Alkoholnarkose. Zweite mittheilung. Ein physikalisch-chemischer. Beitrag zur Theorie der Narcotica, *Arch. Exp. Pathol. Pharmakol.*, 42, 119, 1899.
10. **Overton, E.**, *Studien Uber Die Narkose, Zugleich eion Beitrag zur Allgemeine Pharmakologie*, Fisher, Jenna, 1901.
11. **Traube, J.**, Theorie er Osmose und Narkose, *Arch. Ges. Physiol. (Pflugers)*, 105, 541, 1904.
12. **Fuhner, H.**, Ueber die einwirklung verschiedener alkahole auf die entwirklung der seeigel, *Arch. Exp. Pathol. Pharmakol.*, 52, 69, 1904.
13. **Fuhner, H.**, Pharmakologische studien an see igen leiern; wirkungsgrad der alkohole, *Arch. Exp. Pathol. Pharmakol.*, 52, 69, 1904.
14. **Hansch, C. and Leo, A.**, *Substituent Constant for Correlation Analysis in Chemistry and Biology*, Wiley-Interscience, New York, 1979.
15. **Hansch, C.**, A quantitative approach to biochemical structure-activity relationships, *Acc. Chem. Res.*, 2, 232, 1969.
16. **OECD**, Data Interpretation Guides for Initial Hazard Assessment of Chemicals—Provisional, Organisation for Economic Co-operation and Development, Paris, 1984.
17. **James, A. P. and Martin, A. J. P.**, Gas-Liquid partition chromatography: the separation and microestimation of volatile fatty acids from formic to dodecanoic acid, *Biochem. J.*, 50, 679, 1952.
18. **Woodwell, G. M.**, Toxic substances and ecological cycles, *Sci. Amer.*, 216, 24, 1967.
19. **Mackay, D. and Patterson, S.**, Calculating fugacity, *Environ. Sci. Technol.*, 15, 1006, 1981.

Chapter 2

EVALUATION OF THE BIOCONCENTRATION FACTOR, BIOMAGNIFICATION FACTOR, AND RELATED PHYSICOCHEMICAL PROPERTIES OF ORGANIC COMPOUNDS

Des W. Connell

TABLE OF CONTENTS

I. INTRODUCTION

During the 1950s and 1960s the basic mechanism of bioaccumulation was explained using field data[1] gathered using the newly developed analytical technique of gas chromatography.[2] However, since the early 1970s, following a proposal by Hanelink et al,[3], quantitative measures of the physiochemical properties of the chemical itself have been increasingly used to explain bioaccumulation behavior. Taking a lead from Hansch,[4] researchers studying bioaccumulation began to use the n-octanol-to-water partition coefficient (K_{OW}) as a property which could be used to quantify the bioaccumulation capacity of compounds. A considerable volume of data supporting the value of this physiochemical characteristic has been developed, and this has been reviewed by Connell.[5] The octanol-to-water partition coefficient has considerable advantages over coefficients measured with other possible combinations of phases, since a considerable volume of information has been accumulated on this characteristic in relationship to pharmacological properties. However, in recent years related partition characteristics, such as water solubility and the water to sediment partition coefficient have also been used, as illustrated by Briggs.[6] Typical values for K_{OC} (sediment-to-water partition coefficient in terms of organic carbon) are shown in Table 1, but the use of this characteristic is not discussed further since applications in this area have been limited.

As well as providing an analytical technique for xenobiotic organic compounds, it became apparent in the 1960s that since chromatography was based on partition behavior it could be used to characterize environmental partitioning. Initially, thin layer chromatography was developed as a means of characterizing compounds which may bioaccumulate through using this technique to estimate the K_{OW} value. Later in the 1970s, when high pressure liquid chromatography (HPLC) became readily available, this technique was extensively used to estimate K_{OW} values.

There has been considerable interest in using the molecular structure of a compound to calculate environmental characteristics as well as physicochemical properties. Molecular descriptors which can be utilized are molecular surface area, fragment constants, topological indices, and so on. If these can be applied reliably and accurately, it is clear that considerable savings could be made in the cost of evaluating the environmental properties of new commercial compounds. Lyman et al[7] have comprehensively reviewed many of the various methods employed for chemical property estimation of relevance to environmental behavior of organic compounds.

II. DIRECT CHARACTERIZATION OF THE BIOACCUMULATION CAPACITY OF COMPOUNDS

A. THE BIOCONCENTRATION FACTOR (K_B or BCF)

Connell[5] has described many situations in aquatic systems in which the bioconcentration of organic compounds can be seen to be an equilibrium process involving uptake and loss of a compound between biota and the surrounding water. These situations and others are discussed in Chapter 4 concerning routes for bioconcentration and biomagnification. Under these conditions the Freundlich equation is applicable and the following relationships generally hold

$$C_B = K_B C_W^{1/n}$$

where C_B is concentration in biota, K_B the bioconcentration factor, C_W the concentration in water, and $1/n$ a nonlinearity constant.

The nonlinearity constant allows for deviations from linearity, which often occur at comparatively high concentrations. With most, but not all, environmental situations the xenobiotic organic compounds occur in very low concentrations, and usually the nonlinearity constant is unity ($1/n = 1$). Thus

TABLE 1
Some Physicochemical Parameters and Bioconcentration Factors for Some Typical Compounds Which Bioaccumulate

Compound	Log K_{OW}	Log K_B (fish)	BF (cattle)	Log K_{OC}	S (mg L^{-1})
Chlordane	5.15	4.05	0.3	4.32	0.056
DDT	5.98	4.78	0.9	4.38	0.0017
Dieldrin	5.48	3.76	2.3	4.55	0.022
Lindane	4.82	2.51	0.55	2.96	0.150
Chlorpyrifos	4.99	2.65	0.02	4.13	0.3
Cyhexatin	5.38	2.79	0.0023	>3.64	<1.0
2,4D	1.57	1.30	0.00035	1.30	900
2,4,5T	0.60	1.63	0.0011	1.72	238
TCDD	6.15	4.73	3.5	5.67	0.0002

Adapted from Kenaga, E. E., *Environ. Sci. Technol.*, 14, 553, 1980.

$$C_B = K_B C_W \quad \text{and} \quad K_B = C_B/C_W$$

Thus the bioconcentration factor K_B, or BCF for a compound, is the ratio between C_B in a specified organism or group of organisms, and C_W, as defined by the Freundlich equation operating at low concentrations, where $1/n$ is unity, and where equilibrium has been attained under specified environmental conditions. However, it should be recognized that a variety of terms, now no longer appropriate, such as Bioaccumulation Factor, Accumulation Coefficient, and Concentration Factor, have been previously used to describe this characteristic. Environmental conditions include such aspects as the situation in which the determinations were made, e.g., in aquaria or in the field. In addition, if equilibrium has not been attained due to a limited period of exposure, Hawker and Connell[8] have reported that the K_B values obtained will be lower than the value at equilibrium. This is discussed further in Chapter 6, Section IIIA., Kinetics of Bioconcentration.

It is noteworthy that K_B values are usually not unitless. Measurements of C_B are commonly expressed as mg kg^{-1} (or equivalent units), whereas the units of C_W are usually in mg L^{-1} (or equivalent units), which means that K_B has units of L kg^{-1} as illustrated by the results of Branson et al.,[9] Hamelink,[10] and Oliver and Niimi.[11] Values expressed in this way can be converted into unitless values by mutiplying the K_B value by the density of the biota (kg L^{-1}).

It is also common practice to use the wet weight of the organism to express the C_B value, which is also illustrated by the results of Branson et al.,[9] Hamelink,[10] and Oliver and Niimi.[11] Mackay[12] has concluded that this can lead to variations in the derived K_B values for the same compound due to the differing concentrating capacity of organisms resulting from the presence of different lipid contents within them. This is considered in more detail in Chapter 6, Section II.A. The OECD Guidelines for Testing Chemicals[13] suggest that the lipid or fat content should be measured and C_B should be expressed in terms of lipid. This procedure eliminates this factor as a source of variations in the derived K_B values.

Hamelink et al[3] have conducted an investigation of the behavior of DDT in ecosystems established in artificial and natural pools. This group produced a Freunlich isotherm for the invertebrates to water relationship, which is shown in Figure 1. The regression line has the equation

$$C_B = 13,520 \, C_W + 474 \quad (r = 0.99)$$

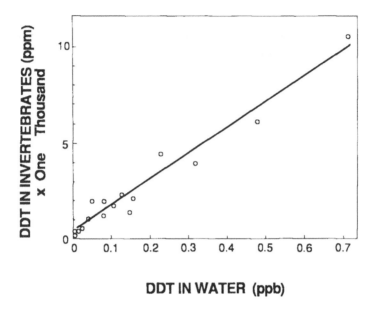

FIGURE 1.   Freundlich-type isotherm for DDT (and metabolites) in the invertebrate-to-water relationship in artificial and natural pools (From Hamelink, J. E., Waybrant, R. C. and Ball, R. C., *Trans. Am. Fish. Soc.*, 100, 207, 1971. Copyright American Fisheries Society. With permission.)

The 474 value is probably due to variations in the system and errors in measurement of the DDT concentrations and can be disregarded. In this equation, the 13,520 is the K_B value for DDT.

Frequently, values of K_B are derived from a limited number of values of C_B and the corresponding C_W, and in some cases, only one pair of values has been used. In the results shown in Figure 1, a single value, in many cases, would have yielded a reasonably accurate result, but in some instances a substantial deviation from the true result would have occurred. This aspect should be kept in mind in evaluating the accuracy of K_B values. Since K_B values cover a wide numerical range, it is convenient to convert them to logarithmic values. A representative set of log K_B values is shown in Table 1.

B. THE BIOMAGNIFICATION FACTOR (BF)

There are situations where the accumulation of xenobiotic compounds cannot be described by a relatively simple equilibrium between an organism and the surrounding water. These have been discussed by Connell,[5] and this situation applies particularly with terrestrial organisms and also some aquatic organisms, and is discussed in Chapters 4 and 7. Nevertheless, the ratio between the concentrations in the organism and the source is used to characterize the bioaccumulation behavior of a compound. When the source is known to be food, the ratio between the concentration of a compound in food C_F, and the consumer C_C, is the Biomagnification Factor (BF). In the literature this factor has been previously described by a variety of terms such as Bioaccumulation Factor, by Garten and Trabalka;[14] Accumulation Factor, by Sugiura et al.;[15] Bioconcentration Factor, by Geyer et al.;[16] and Concentration Factor, by Canton et al.[17]

The Biomagnification Factor is defined as the ratio between C_F and C_C, i.e., C_F/C_C, for a specified food type and specific consumer or group of consumers over a designated time period and under specified environmental conditions. Some values for BF for different compounds are shown in Table 1. In contrast to the values for K_B, the BF values are usually relatively small, with TCDD, the maximum in Table 1, having a value of 3.5. Usually,

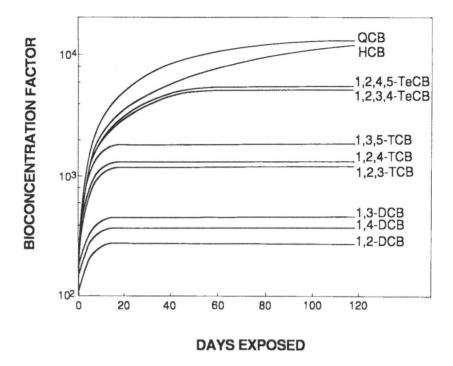

FIGURE 2. Bioconcentration of chlorobenzenes from water over time by fish. (From Oliver, B. G. and Niimi, A. J., *Environ. Sci. Technol.*, 17, 287, 1983. Copyright American Chemical Society. With permission.)

these values can be conveniently expressed without the use of logarithms, as has been done with the data in Table 1.

Although the theoretical basis for this factor has not been thoroughly evaluated, the evidence suggests that possibly an equilibrium between stomach contents and body tissues is involved. Kenaga[18] has shown that the BF values for various compounds taken up by cattle and swine are correlated with the K_B values for fish. This suggests that processes somewhat akin to those in the fish-to-water system are operating. However, in more recent work, Garten and Trabalka[14] have found a poor relationship between K_B for fish and BF values for a variety of terrestrial animals, with a range of different compounds. Under these circumstances, it is inappropriate to use the common designation for equilibrium partition coefficients, K, to describe this characteristic, and the term BF is more appropriate.

C. EXPERIMENTAL PROCEDURES

The experimental determination of K_B values by any procedure is valid only when the test compound is fully dissolved within the water. The behavior patterns of compounds present in water as small droplets or aggregates of molecules is not clear, but does not have the characteristics previously outlined for bioconcentration. A common method for the experimental determination of K_B involves exposure of the organism to the compound dissolved in water (C_W), and the determination of the concentration present in the biota (C_B) by periodic analysis. When the C_B value ceases to increase with time, equilibrium is considered to have been attained, and the K_B value is calculated as C_B/C_W. This method is referred to as the "plateau" or "steady state" method, and is illustrated by the results obtained by Oliver and Niimi[19] shown in Figure 2. In this experiment, the K_B values at 119 d were taken to represent the equilibrium K_B values. However, more reliable results can be obtained if this procedure is replicated at different concentrations to enable a number of

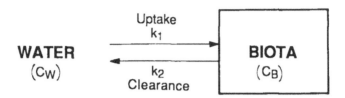

FIGURE 3.   Diagrammatic illustration of the biota to water transfers of a xenobiotic compound where the biota are a single compartment.

pairs of C_W and the corresponding C_B results to be obtained. A regression line fitted to this data yields a sorption isotherm similar to a Freundlich isotherm, and the slope of the line is equal to K_B.

It can be seen that some compounds in Figure 2 were continuing to increase in K_B values at 119 d, and thus had not reached equilibrium. In more recent work, Hawker and Connell[8] have shown that lengthy periods are required to attain equilibrium with many higher molecular weight compounds. As a result of difficulties in achieving equilibrium and the lengthy time periods required for experiments, Branson et al.[9] have proposed an alternative short-term test to measure K_B values. This approach assumes that the water to biota bioconcentration process can be represented by a single compartment model for the biota, as illustrated in Figure 3. In this system,

Rate of Bioconcentration = Rate of Uptake — Rate of Clearance

thus
$$\frac{dC_B}{dt} = k_1 C_W - k_2 C_B$$

where C_W and C_B are the concentrations in water and biota, respectively, and k_1 and k_2 the uptake and clearance rate constants, respectively.

When equilibrium is reached C_B ceases to increase, and the rate of bioconcentration dC_B/dt is zero, then

$$C_B/C_W = k_1/k_2 = K_B$$

This means that instead of continuing an experiment until equilibrium is reached, the experiment need only be continued for a sufficient period of time to allow the determination of k_1 and k_2, and K_B can be calculated from these values. This method of determining the bioconcentration factor is referred to as the ''kinetic'' method.

The kinetic method to measure K_B requires a modified experimental procedure, as compared to that used with the plateau method. The uptake curve is obtained by a similar exposure procedure to that used with the plateau method, but after a sufficient period has elapsed the animals are transferred to water which is free of the xenobiotic chemical. The loss of the chemical in the biota is now measured over time, allowing k_2 to be calculated, as well as k_1 from the initial uptake curve. This procedure is illustrated by the results obtained by Branson et al.,[9] which are shown in Figure 4.

Clearly, the kinetic method is dependent on the validity of the single compartment model, and is based on measurements made only in the initial stages of uptake and clearance. More extended time periods of uptake and clearance may reveal deviations from the values which were obtained in the initial stages, as has been reviewed by Davie and Dobbs.[20] These authors report data available for comparison of K_B values obtained by both the kinetic and plateau methods as shown in Table 2. These results suggest that the K_B values obtained by the plateau method may exceed those obtained by the kinetic method, although more results

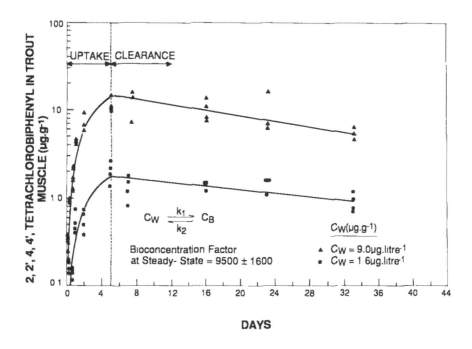

FIGURE 4. Concentrations 2, 2′, 4, 4′-tetrachlorobiphenyl in rainbow trout muscle during uptake from water and clearance after transfer of the fish to uncontaminated water. (From Branson, D. R., Blau, G. E., Alexander, H. C. and Neely, W. B., *Trans. Am. Fish Soc.*, 104, 785, 1975. Copyright American Fisheries Society. With permission.)

TABLE 2
Comparison of K_B Values Obtained by the Kinetic and Steady State[a] Methods

Compound	Kinetic K_B	Steady state K_B
DDT	52,358	100,000
Hexachlorobenzene	7880	18,600
Tetradecylheptaethoxylate	850	700
Sodium dodecylbenzene sulfonate	286	20
1,4-Dichlorobenzene	215	60
Diphenyl oxide	190	470
Tetrachloroethylene	39.6	49
Carbon tetrachloride	17.7	30

[a] Also referred to as the "plateau" method.

From Davies, R. P. and Dobbs, A. J., *Water Res.*, 18, 1253, 1984. Copyright Pergamon Press. With permission.

are needed to reach a definite conclusion on this aspect. Some examples of the use of both methods to obtain K_B values are outlined in Table 3.

Summaries of examples of the variety of experimental procedures which have been used to obtain K_B values with fish are shown in Table 3. The methods used by Neely et al.,[21] Branson et al.,[9] Bishop and Maki,[22] Sugiura et al.,[23] Southward et al.,[24] and Muir et al.[25] involved relatively small laboratory aquaria and correspondingly small fish. In contrast, Oliver and Niimi[19] have utilized much larger laboratory-based tanks and fish. Also, Hamelink et al.[3] conducted experiments on a scale approaching conditions in natural aquatic systems.

TABLE 3
Examples of the Procedures Used to Determine K_B with Fish

Test organisms	Test environment	Test compound and addition technique	Monitoring of test compound concentrations	Calculation of K_B	Ref.
Common shiner (*Notropis cornutus*), northern red belly and dace (*Chrosomus cos*), young large mouth bass (*Micropterus salmsides*), green sunfish (*Lepomis cyanellus*), pumpkin seed sunfish (*Lepomis gibbosus*)	Artificial and natural pools	DDT; single addition in solution to water or on sand, added to bottom sediments	Periodic analysis of water and organisms	Slope of regression line of C_B versus C_W at equilibrium	3
Rainbow trout (*Salmo gairdneri*) (8—10 g)	Laboratory aquaria (12 L)	Chlorinated hydrocarbons and other compounds; proportional diluter giving flow-through followed by transfer of fish to clean water	Periodic analysis of water and organisms during uptake and clearance	Calculation of k_1 and k_2 then K_B from k_1/k_2	21
Rainbow trout (*Salmo gairdneri*) (8—10 g)	Laboratory aquaria (12 L)	2,2′,4,4′-tetrachlorobiphenyl; proportional diluter giving flow-through followed by transfer of fish to clean water	Periodic analysis of water and organisms during uptake and clearance	Calculation of k_1 and k_2 then K_B from k_1/k_2	9
Juvenile bluegill (*Lepomis macrochirus*) (av. wt. 0.49 g)	Laboratory aquaria (37 L)	EDTA, LAS, AE and DET; proportional diluter as above followed by transfer of fish to clean water	Periodic analysis of water and organism during uptake and clearance and to equilibrium	Calculation of K_B from k_1/k_2 C_B/C_W at equilibrium	22
Killfish (*Cryzias latipes*)	Laboratory aquaria (13 L)	PCBs and PBBs, flow-through of standard solutions of compounds	Periodic analysis up to 20 days	C_B/C_W at 20 d	23
Fathead minnows (*Pimephales promelas*) (av wt: 75 mg and 140 mg)	Laboratory aquaria (8 L)	Azaarenes; single addition to water	Periodic analysis	k_1/k_2	24
Rainbow trout (*Salmo gairdneri*) (av. wt: 250 g)	Tanks (1 × 1 × 0.5 m)	Chlorobenzenes; water flow-through with continuous introduction of compound in solution	Periodic analysis up to 119 d	C_B/C_W at 119 d	19
Rainbow trout (*Salmo gairdneri*) (fry 0.1-	Laboratory	1,3,6,8-TCDD and OCDD;	Periodic analysis dur-	k_1/k_2	25

0.5 g), fathead minnow (*Pimephales promelas*) (1.5—2.5 g)	aquaria (15 L)	aqueous solutions produced using generator columns followed by dilution in a flow-through system followed by transfer of fish to clean water	ing uptake and clearance		
Catfish (*Ictalurus melas*) (1—2 g)	Laboratory aquaria	Various pesticides; single and four-fold applications	Periodic analysis during uptake and clearance	Slope of regression line of C_B versus C_w at equilibrium	26

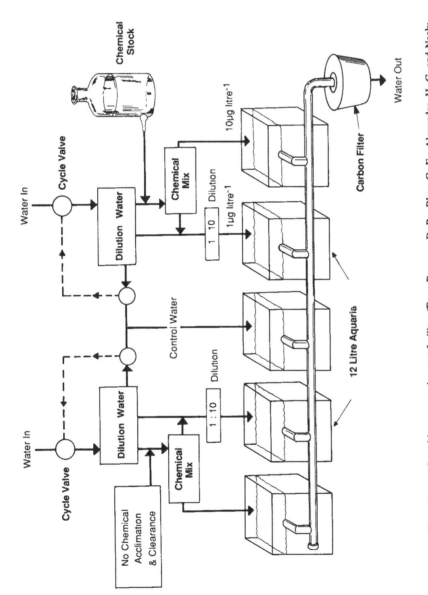

FIGURE 5.   Flow diagram for a bioconcentration test facility. (From Branson, D. R., Blau, G. E., Alexander, H. C. and Neely, W. B., *Trans. Am. Fish. Soc.*, 104, 785, 1975. Copyright American Fisheries Society. With permission.)

The system used to control water concentrations of the test chemicals is particularly important. The early experiments of Hamelink et al.[3] involved the addition of an initial amount of chemical to water, or on sediments, and following the changes in water concentrations over time by analysis. Somewhat similarly, Southworth et al.[24] conducted experiments using a single initial addition of chemical to the water. This method is usually referred to as "static" testing. If depletion of the compound in the water occurs rapidly or before effective equilibrium is reached, the results obtained using this method will not be accurate. However, the losses of compound in the water can be compensated for by changing the aquarium water on several occasions and replacing it with water containing the chemical at the required concentration. This procedure is illustrated by the experiments conducted by Ellegehausen et al.[26]

Another exposure method, the "flow-through" procedure, is often used. This involves a controlled continuous addition of compound, usually in solution, as well as the flow-through of fresh uncontaminated water. Branson et al.,[9] Oliver and Niimi,[19] Neely et al.,[21] Bishop and Maki,[22] Sugiura et al.,[23] and Muir et al.[25] have used this procedure. The procedure has the advantage that water concentrations are controlled, and so experiments can be conducted over comparatively long time periods until equilibrium is attained. This would be expected to give more accurate and reproducible results. A diagrammatic representation of a system for controlling the concentrations in the water and providing for flow-through of water is shown in Figure 5.

In some situations, particularly with compounds which are very insoluble in water, a column filled with the compound dispersed on a solid inert packing described as a "generator" column is used.[25] Water is initially circulated through the generator column where equilibrium is established, and the water concentration approaches the water solubility value for the test compound. This water may be diluted, or used as produced, in an aquarium flow-through system.

The OECD Guidelines for Testing Chemicals[13] includes comprehensive descriptions of several methods to determine K_B values with fish. The methods described involve both static and flow-through techniques, and K_B values are calculated by the plateau and kinetic methods.

Most bioconcentration experiments have been carried out on fish, but a variety of other organisms have been subject to this testing procedure. In principle, the methods used with these other organisms are similar to those used with fish. For example, with comparatively small organisms, such as zooplankton and algae, static experiments are often conducted in realtively small aquaria or glass vessels. Examples of this procedure have been provided by Geyer et al.[27] with *Chlorella*, Harding and Vass[28] with marine planktonic crustacea, Crosby and Tucker[29] with *Daphnia magna*, and Southward et al.[30] with *Daphnia pulix*. In a limited number of experiments with these organisms, flow-through techniques were used, as was done by Sondergren.[31] Baughman and Paris[32] have reviewed much of this information. Similarly, mussels and other bivalves have been used in a variety of experiments based on the same principles. Vreeland,[33] Solbakken et al.[34] and Lee et al.[35] have used the static procedure, whereas Ernst[36] used a static procedure with periodic dosing with the test chemical, and Renberg et al.[37] used the flow-through method.

With the experimental determination of BF values (Biomagnification Factors) a more complex situation exists. In the bioconcentration process from water, the water is relatively consistent in physicochemical properties. However, with biomagnification of xenobiotic compounds from food there are many possible types of food which vary considerably in their physicochemical properties. This is illustrated by the data in Table 4. The food types used differ considerably in physical form and a variety of other potentially important factors, such as fat content. In some other cases, illustrated by the procedures used by Metcalf et al.,[41] an attempt was made to simulate a multiple component food chain in the laboratory. Thus, overall there is not a consistent pattern of experimental procedures used to measure this characteristic.

TABLE 4
Some Examples of the Procedures Used to Determine BF

Substance	Food	Consumer	Experimental method	Ref.
HOED	Tuberficid worms	Reticulate sculpin	HOED in worms was obtained by controlled exposure and fed to fish over 14 d	38
Organochlorines	Commercial trout chow	Yearling coho salmon	Added to food and fed to fish over 27 d	39
Dieldrin	*Daphnia magna*	Guppies	Dieldrin in *D. magna* was obtained by controlled exposure and fed to fish over 30 d	40
DDT, DDE, DDD, methoxychlor	*Sorghum Extigmene* (excreta) *Oedogonium Estigmene* diatoms plankton *Culex*	*Estigmene Oedogonium Physa* diatoms plankton *Culex Gambusia*	Laboratory model ecosystem with compounds added to *Sorghum*	41
PCBs	Commercial fish food	Goldfish	PCBs added to food fed to fish over 150 days	42
TCDD	Normal diet	Rats, cattle,	TCDD added to the diet fed to biota	43

III. PROPERTIES USED TO MEASURE THE PARTITIONING CHARACTERISTICS OF COMPOUNDS

A. THE OCTANOL TO WATER PARTITION COEFFICIENT (K_{ow}, P_{ow}, P)
1. Definition and Nature

In the classic work by Overton and Meyer outlined in Chapter 1, Introduction, it was recognized that xenobiotic chemicals could induce physiological effects in organisms which were related to the oil (or lipoid substance)-to-water partition coefficient. The oil-to-water partition concept was based on earlier work on partitioning behavior between two immiscible liquids by Nernst[44] and other workers, as described in the Introduction. Nernst[44] found that there was a constant ratio between the concentration in the lipoid solvent and the concentration in the water at equilibrium and at constant temperature. This was defined as the Distribution Ratio and later described as the Partition Coefficient. This means that $K_{LW} = C_L/C_W$, where K_{LW} is the partition coefficient between lipoid solvent and water, and C_L, the concentration in the lipoid solvent, and C_W the concentration in water. In thermodynamic terms this equilibrium is characterized by equality of the chemical potentials of the solute in the two immiscible solvents. Thus

$$\mu_L = \mu_L^o + RT \ln C_L$$

and

$$\mu_W = \mu_W^o + RT \ln C_W$$

where μ_L and μ_W are the chemical potentials of the solute in the lipoid solvent and water, respectively, μ_L^o and μ_W^o the chemical potentials when the concentrations of the solute are at 0, R the Universal Gas Constant, and T the absolute temperature.

These equations are more precise when activities are used instead of concentrations. However, with the dilute solutions, which are of principle concern regarding xenobiotic chemicals in the environment, these properties are of approximately equal value.

Thus at equilibrium

$$K_{LW} = C_L/C_W = \exp[-(\mu_L^o - \mu_W^o)/RT] = \text{constant} \qquad (1)$$

Thus the partition coefficient is constant under defined conditions. In addition, Nernst[44] has also derived an expression for K_{LW} using the Henry's Law Constants. The solute activity and its fugacity at high dilution are proportional to its concentration, and these properties are equal when the solute is at equilibrium between the two phases. Thus

$$H_L X_L = H_W X_W$$

where H_L and H_W are the Henry's Law Constants of the solute in lipoid solvent and water, respectively, and X_L and X_W the mole fractions in the water saturated lipoid solvent and the lipoid solvent saturated with water, respectively. But at dilute concentrations, $X = C\overline{V}$, where X is the mole fraction of the solute, C the molar concentration, and $\overline{V}$ the molar volume of the phase corrected for mutual solvent water saturation. The

$$H_L C_L \overline{V}_L = H_W C_W \overline{V}_W$$

and

$$C_L/C_W = H_W \overline{V}_W / H_L \overline{V}_L = K_{LW}$$

In many studies the lipid component of the two immiscible solvents to measure partition coefficients was a natural oil, such as olive oil, although many other substances have been used. In 1971 Leo et al.[45] reviewed the information on partition coefficients then available, and later Hansch and Dunn[46] suggested that n-ocantol and water were the most appropriate solvent pair to standardize results.

Several reasons were put forward for the selection of n-octanol as the lipoid component of the system. The principle reason is that it was believed that this substance provided a reasonable model of the macro molecular system in which xenobiotic compounds operate within organisms. Also, n-octanol has the desirable property of a poor capacity to dissolve water, and it has somewhat similar polarity to the fats which occur in most organisms. The presence of a hydroxyl group gives it the capacity to act as a hydrogen bond donor as well as an acceptor. Being a liquid is also an advantage, since in the experimental determinations for the measurement of the partition coefficient it could be expected that a liquid would attain equilibrium more rapidly and more reproducibly than a solid. Another advantage of n-octanol is that it is a single compound rather than a mixture, as with natural fats and oils, and can be purified to a high degree, and so phases of consistent properties can be used for partition coefficient determinations.

The n-octanol to water partition coefficient (K_{OW}) is defined as

$$K_{OW} = C_O/C_W$$

where C_O and C_W are the concentrations of solute in the n-octanol and water, respectively, at equilibrium and at a constant temperature.

In this equation K_{OW} is dimensionless, since C_W and C_O are both measured in the same units of mass per unit volume. K_{OW} is usually measured at 20 or 25°C, but is only slightly temperature dependent, as suggested by Equation (1). Lyman[47] has reported this dependence to be 0.001 to 0.01 K_{OW} units per degree, and may be either positive or negative depending on the solute. It also has a slight dependence on concentration, as reported by Mackay.[48] Lyman[47] found that the measured values for a variety of compounds ranged from 10^{-3} to 10^7, but it is more convenient to express these values on a logarithmic scale using the base 10, i.e., log K_{OW} ranges from -3 to 7. Some examples of log K_{OW} values are shown in Table 1. The abbreviation K_{OW} is used throughout this book, but P_{OW} and P are also in common use.

For a compound to exhibit the partition behavior described previously, it must be in the same state in both the octanol and the water. Although most of the environmental compounds of interest are nonelectrolytes, some are weak electrolytes, and the proportion of the compound present in the ionized and unionized form will vary with pH. As a result, a variation in K_{OW} values will result from measurements taken at different pH values. Thus, to obtain consistent and accurate values, the pH and the presence of other electrolytes, which will also influence the degree of ionization, must be controlled. This is discussed further in Chapter 3, General Characteristics of Substances Which Exhibit Bioaccumulation.

The K_{OW} values are the major partitioning property currently used in the investigation of quantitative structure-activity relationships with bioaccumulation. One of the reasons for this is the large volume of data available on K_{OW} values for a wide range of compounds. Chiou[49] has provided evidence for the validity of K_{OW} as a predictor of bioaccumulation. He found that log K_{OW} was closely correlated with the triolein-to-water partition coefficient where triolein was seen as a model for organism lipid. Over the log K_{OW} range of one to about five a close correlation was obtained, but increasing deviations from this relationship are apparent at log K_{OW} values which are greater than five.

2. Experimental Measurement

If the solute has reasonable solubility in both octanol and water, then relatively simple methods can be used to experimentally determine its K_{OW} value. Octanol and water are placed in a glass vessel at constant temperature with the solute and stirred or shaken until equilibrium is established. This method of determination is often referred to as the "shake flask" method. The establishment of equilibrium can be checked by periodic withdrawal and analysis of samples. The K_{OW} value is determined from the concentrations in octanol and water at equilibrium. Hansch et al.[50] and Fugita et al.[51] have used this method in some of their investigations, and James[52] has described the various methods used based on this principle. The OCED Guidelines for Testing of Chemicals[13] contains detailed experimental procedures for using the "shake flask" method.

The principle of the "shake flask" technique has been extended by mechanically mixing the two phases and drawing off the mixture through a special centrifuge, then monitoring the composition of the separated phases until equilibrium is reached. This is often referred to as the AKUFVE System, and has been described by James[52] together with a variety of related methods.

These methods have a number of disadvantages and limitations, particularly with respect to many environmental xenobiotic compounds. These compounds are very hydrophobic, and the separation of the phases can be incomplete. Some of the compound can be retained in the aqueous phase as very small aggregates of molecules, often termed micelles, which are difficult to remove. Although the quantity may be small, the presence of these, or compounds adsorbed onto any colloidal matter present, leads to very large errors in the water concentration, since it is extremely low. This results in a consequent large error in the K_{OW} value obtained. The analytical detection of the extremely low concentrations of very hydrophobic compounds present in water is often difficult, and in some cases impossible, with current analytical methodology.

Some of the problems described above have been overcome by using the mechanically mixed system with continuous monitoring by pumping both phases through analytical instruments. In these situations, selective hydrophobic and hydrophilic filters are used as probes in the mixture to withdraw the separate phases. An example of a system designed by Tomlinson et al.[53] is shown in Figure 6.

In recent years, a different type of two phase system has been designed to achieve equilibrium partitioning, and to overcome many of the problems previously described. This system is usually referred to as a "generator column", and has been described by De Voe et al.,[54] Wasik et al.,[55] and Miller et al.[56] In essence, generator columns are tubes maintained at a fixed temperature and filled with an inert dispersed material, such as glass beads, Chromosorb, or similar material onto which the solute in octanol solution is coated. Water is slowly passed through the column, allowing equilibrium to be established. This system provides a large surface area for interchange between octanol and water without the mechanical dispersal of one phase within the other. The procedure used by Woodburn et al.[57] is typical of these methods described for K_{OW} value determination and is outlined below:

1. The solute was dissolved in octanol at a concentration of about 1 percent W/W solution.
2. This solution was equilibrated with water.
3. The concentration of solute in this equilibrium solution of octanol was then accurately determined.
4. This equilibrated octanol solution was then applied to the generator column packed with inert material.
5. Water saturated with octanol was passed slowly through the column and collected.
6. The concentration of solute in the water was determined by an appropriate method.
7. The K_{OW} value was determined by dividing the concentration in the original octanol solution by the concentration in the collected water.

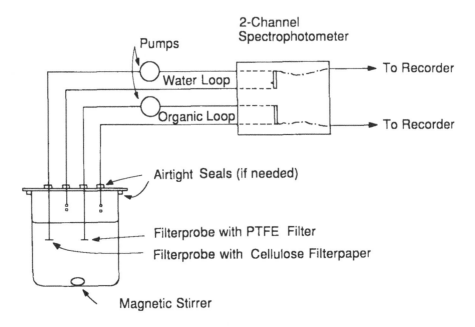

FIGURE 6.   Schematic diagram of a typical filter probe system. (From Tomlinson, E., Davis, S. S., Parr, G. D., James, M., Farraj, N., Kinkel, J. F. M., Gaisser, D., and Wynne, H. J., *Partition Coefficient — Determination and Estimation,* Pergamon, Elmsford, NY, 1986, 83. Copyright Pergamon Press. With permission.)

Wasik et al.[55] have found good agreement between K_{OW} determined by this method and by other methods. They estimated the error to be 4% relative to the mean value of K_{OW}. These authors have outlined the advantages of the generator column method over the shake flask method previously described. The principle advantage is that the rate of water flow is set slow enough to avoid colloidal dispersion and the formation of aggregates and micelles. In addition, the interphase surface area is sufficiently large to allow equilibrium to be rapidly established. In this continuous flow system the generator column tube walls become equilibrated with the rest of the system, and so errors due to wall adsorption are minimized.

In conjunction with the generator column technique chromatographic methods are used, but only as analytical tools for the analysis of the separated phases. However, chromatographic techniques are based on the principle of a large number of partitions between two immiscible phases. Thus, a relationship between chromatographic characteristics and K_{OW} would be expected. The accuracy of these relationships would be expected to depend on how closely the two chromatographic phases resemble the octanol and water system.

A number of chromatographic techniques have been used to measure K_{OW} values and have been described by James.[52] In thin layer chromatography (TLC) and paper chromatography the chromatographic behavior of a compound is characterized as the R_F value (the distance the compound moves from the origin/the distance the solvent moves from the origin). The R_F values can be used to calculate the parameter, R_M

$$R_M = \log(1/R_F - 1)$$

The R_M value can be used to calculate log K_{OW} values from the equation

$$\log K_{OW} = R_M + \text{Constant}$$

For this expression to give useful results the stationary and mobile phases should resemble octanol and water. Usually, the stationary phase in these chromatographic techniques is hydrophilic and the mobile phase is hydrophobic. This means that the nature of these phases must be reversed (referred to as reversed phase chromatography) for this application to be successful. For example, Boyce and Milborrow[58] used TLC in which the silica gel phase was impregnated with paraffin, and acetone in water solutions were used as the mobile phase. Tomlinson[59] has used a reversed phase TLC system to determine log K_{OW} values. Somewhat similar methods have been used by Bowen et al.[60] with paper chromatography to obtain log K_{OW} valves. Both TLC and paper chromatography suffer from a lack of precision and accuracy with which R_F values can be obtained and a consequent effect on the accuracy of the log K_{OW} values.

Gas liquid chromatography has been used to determine log K_{OW} values as reviewed by James.[52] The retention time (t_R) of a compound through a gas liquid chromatography column is represented by

$$t_R = t_0(1 + qK_{OW})$$

where t_0 is the retention time of a compound which has no interaction with the liquid phase, and q the ratio of the stationary and mobile phases.

By rearrangement of this equation the following is obtained

$$\log K_{OW} = \log(t_R/t_0 - 1) + \log 1/q$$

By using reference compounds of known log K_{OW} values, the values of unknowns can be obtained through the use of this equation. The t_R values of compounds can be obtained with a high level of precision, but the use of elevated temperatures for the gas chromatographic process to determine log K_{OW} values to be applied at 20° or 25°C may introduce errors. Also the differences between the octanol to water system and the gas to liquid system may cause deviations from the true values.

High pressure or high performance liquid chromatography (HPLC) is the chromatographic system which can be arranged to resemble the octanol-to-water partitioning system most closely. For example, Mirrlees et al.[61] chromatographed solutes with octanol saturated with water as the mobile phase, and water saturated with octanol entrained on an inert support as the stationary phase. Weber et al.[62] have outlined the theory of the use of HPLC to determine the log K_{OW} of a compound. The partition coefficient of a compound between the stationary and mobile phases (K_P) is

$$K_P = C_i^s/C_i^m$$

where C_i^s and C_i^m are the concentrations of a compound in the stationary phase and mobile phases, respectively.

Since $C_i^s = n_i^s/V_s$ and $C_i^m = n_i^m/V_m$ where n_i^s and n_i^m are the numbers of moles in the stationary and mobile phases, respectively, and V_s and V_m are the volumes of the stationary and mobile phases, respectively. Thus

$$K_P = \frac{n_i^s V_m}{n_i^m V_s}$$

But the capacity factor, or k', is the equilibrium ratio of the number of moles in the stationary and mobile phases.

$$K_P = k^1(V_m/V_s) \tag{2}$$

The volume of solvent required to elute a compound, V_t, increases as K_P increases, and is

K_p times the V_s. But V_t should be corrected for the volume of the mobile phase, V_m, in the HPLC column because this volume plays no quantitative role in the chromatographic separation. For example, if a compound has no interaction whatever with the stationary phase it would require a volume size of V_m to move it through the chromatographic column. Thus $K_p = (V_t - V_m)/V_s$ and by combination of this equation with Equation (2) it can be shown that

$$k' = (V_t - V_m)/V_m$$

This can be related to retention times by dividing through by the flow rate of the mobile phase. Therefore

$$k' = (t_R - t_0)/t_0$$

where t_R is the retention time of a compound, and t_0 the retention of a compound which has no interaction with the stationary phase. This latter value can also be seen as the retention time of the mobile phase itself. Thus

$$K_p = k'(V_m/V_s) = (V_m/V_s)(t_R - t_0)/t_0$$

But V_m/V_s and t_0 are constant in any particular HPLC separation, and if the mobile phase and stationary phases are octanol and water, then $K_p = K_{ow}$. Since the logarithms provide a more convenient scale, then

$$\log K_{ow} = \log k^1 + \text{constant} = \log t_c + \text{constant} \tag{3}$$

where t_c is the corrected retention time, i.e., $t_R - t_0$.

Mirrlees et al.[61] found the octanol and water HPLC system was satisfactory for determining log K_{ow} values from -0.3 to $+3.7$. They obtained values for the constants in the equations above by chromatographing solutes of known log K_{ow} values. Compounds with log K_{ow} values greater than 3.7 had retention times too long to be accurately and conveniently measured. Many of the xenobiotic compounds of environmental interest have log K_{ow} values greater than 3.7. To extend the range of the HPLC system, different stationary and mobile phases are necessary. The HPLC system used usually involves the use of a reversed phase having a chemically or mechanically bound nonpolar phase assiociated with inert support, and a water and organic solvent, such as methanol or acetonitrile, as mobile phase. With this system the phases are not directly analogous to the octanol-to-water system, and the K_p to K_{ow} relationship may not be direct and linear, and Equation 3 may take the form

$$\log K_{ow} = \text{constant} \log k^1 + \text{constant} = \text{constant} \log t_c + \text{constant}$$

The values of the constants are obtained by calibration with compounds having known log K_{ow} values.

The method described above has proved to be a good technique to measure log K_{ow} values for a wide variety of compounds. For example, McCall[63] and Henry et al.[64] have used it successfully with certain pharmaceuticals, whereas Swann et al.[65] and Ellgehausen et al.[66] have used it with environmental xenobiotic compounds. Figure 7 produced by Chin et al.[67] illustrates the type of results usually obtained. However, Brooke et al.[68] measured the retention characteristics of a set of compounds using several different concentrations of aqueous methanol as the mobile phase. Retention volumes were then extrapolated to a mobile phase concentration of 100% water. They found this gave an improved correlation with reliably known log K_{ow} values from the literature. But they suggested that the HPLC

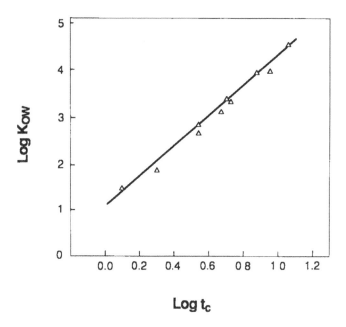

FIGURE 7. Correlation between the log K_{ow} values and corrected retention times (t_c) on HPLC for various compounds. (From Chin, Y. P., Weber, W. J. and Voice, T. C., *Water Res.*, 20, 1443, 1986. Copyright Pergamon Press. With permission.)

procedure does not always provide reliable results. It is interesting to note the investigations of K_{ow} values of PCB by Bruggeman et al.[69] and Rapport and Eisenreich,[70] which indicated deviations from the expected relationships.

3. Calculation of K_{ow}

A wide variety of methods, which can often be obtained from the literature, are available for calculating K_{ow} values from molecular descriptors and other characteristics. Lyman[71] and Rekker[72] have thoroughly reviewed the methods used in many of these calculations and these are summarized in Table 5. A brief review of some of these methods is outlined below. Only the methods based on fragment constants and structural factors, regression equations based on partition coefficient between lipoid solvent and water (K_{LW}), and regression equations based on water solubility, are considered below since these provide the most readily applicable and accurate techniques.

a. Preliminary Estimation

A preliminary estimate of K_{ow} can be obtained from solubility data as described in the OECD Guidelines for Testing Chemicals[13] and discussed by Mackay.[49] The solubility of a compound in a solvent can be expressed as

$$S = (X_w M)/V_w$$

where S is the solubility in water in units of mass per unit volume, X_w the mole fraction of solute in water, M the molecular weight of solute, and V_w the molar volume of water.

Similarly the solubility in octanol can be represented by

$$S_o = (X_o M)/V_o$$

thus

TABLE 5
Overview of Estimation Methods for K_{ow}

Basis for method	Information required[a]	Comments
Fragment constants and structural factors (substituent constants, fragmental constants, parachor, molecular connectivity)	Structure (K_{ow} for structurally related compound)[b]	Fairly accurate Wide range of applicability Requires some practice
Regression equations (Collanders equation)	K_{Lw}	Easy, rapid calculations Limited applicability Fairly accurate
Regression equations (Water solubility)	S	Easy, rapid calculations Wide range of applicability Less accurate
The following two approaches are not recommended[c]		
Regression equations	K_{oc}	Relatively large method error
Regression equations	K_B	Relatively large method error
A Method for the venturesome (not covered in this book)		
Uses estimated activity coefficients	Structure	Calculations are lengthy and difficult Limited applicability for functional groups Fairly accurate

[a] K_{ow} = octanol/water partition coefficient; K_{Lw} = organic solvent/water partition coefficient; S = water solubility; K_{oc} = soil adsorption coefficient based on organic carbon; K_B = bioconcentration factors for aquatic life.

[b] Only required with substituent constants method.

[c] There methods involve relatively large method errors because values of K_{oc} and K_B are highly variable.

Adapted from Lyman, W. J., *Handbook of Chemical Property Estimation Methods*, McGraw-Hill, New York, 1982, 1.

$$S_o/S = X_o V_w / X_w V_o$$

If the octanol and water are in equilibrium in terms of the solute, but not considering equilibrium between the water and octanol phases, then

$$X_o Y_o = X_w Y_w$$

where v_o and v_w are the activity coefficients of the solute in the pure solvents. Now

$$S_o/S = Y_w V_w / Y_o V_o = K_{ow}$$

This means that the K_{ow} value is the ratio of the solubilities in octanol and water. Lyman[71] reports that this method does not take into account the mutual solubility of octanol and water and the OECD Guidelines for Testing Chemicals[13] note that the reliability of this method decreases with increasing structural complexity of solute.

b. Substituent Constants

Within an homologous series a consistent additive effect is observed with each increase in CH_2. For example, Coates et al.[73] in investigations of the n-alkanes found

$$\log K_{ow} = 0.535 \, N_c - 0.302 \qquad (r = 0.999)$$

where N_c is the number of carbon atoms in the molecule.

TABLE 6
Some π Values for Some Common Functional Groups

Function	Aromatic	Aliphatic
NH_2	−1.23	−1.19
I	1.12	1.00
$S-CH_3$	0.61	0.45
$COCH_3$	−0.55	−0.71
$CONH_2$	−1.49	−1.71
$COOCH_3$	−0.01	−0.27
Br	0.86 (ortho)	0.60
Br	1.12 (para)	—
CN	−0.57	−0.84
F	0.14	−0.17
Cl	0.76 (ortho)	0.39
Cl	1.04 (meta)	—
Cl	0.98 (para)	—
COOH	−0.28	−0.67
OCH_3	−0.02	−0.47
OC_6H_5	2.08	1.61
$N(CH_3)_2$	0.18	−0.30
NO_2	−0.28	−0.85
CH_3	0.84 (ortho)	—
CH_3	0.60 (para)	—
CH_3	0.54 (meta)	—
CH	−0.31 (para)	−1.16
C_6H_6	2.13	—
C_6H_5OH	1.46	—
$-CH_2-$	—	0.50
CF_3	1.07 (para)	—

From Neely, W. B., *Chemicals in the Environment*, Marcel Dekker, New York, 1980, 72. Copyright Marcel Dekker. With permission.

Thus, in this case the consistent increment is 0.535 for each carbon atom or $^{CH}2$ grouping. Hansch and co-workers established this relationship in a general form, as described by Fujita et al.[51] and Leo et al.[45] and defined the substituent constant (π_x) as the log K_{ow} due to the functional group X added to a parent molecule. Thus

$$\pi_x = \log P_x - \log P_H$$

where P_H is the parent molecule and P_x the derivative of P_H with a substituent X.

Thus, for example, π_{Cl} can be calculated as

$$\pi_{Cl} = \log K_{ow} \text{ (chlorobenzene)} - \log K_{ow} \text{ (benzene)}$$

A set of π values for common groups reported by Neely[74] are shown in Table 6. To calculate log K_{ow} for a compound using these values, the parent structure and its log K_{ow} must be known. The log K_{ow} is calculated by addition of the log K_{ow} of the parent structure and the π values of the substituents. Thus

$$\log P_X = \sum_i \pi_{Xi} + \log P_H$$

An example of using this method for calculating the log K_{ow} for chlorpyrifos is shown in

TABLE 7
Calculation of the Log K_{ow} Value of Chlorpyrifos

Structure of Chlorpynfos

1. $\log K_{ow}$ of $-O-\overset{\overset{\displaystyle S}{\|}}{P}(OC_2H_5)_2$

   $\log K_{ow}$ of benzene = 3.46

   $\log K_{ow}$ of $\langle O \rangle - O\overset{\overset{\displaystyle S}{\|}}{P}(OC_2H_5)_2$ = 2.13

   thus $\log k_{ow}$ of $-O-\overset{\overset{\displaystyle S}{\|}}{P}(OC_2H_5)_2$ = 3.46 − 2.13

   = 1.33

2. $\log K_{ow}$ of pyridine   = 0.66

3. $\log K_{ow}$ of chlorines

   $\log K_{ow}$ of chlorines   = $\log K_{ow}$ · Clortho

   + 2 × $\log K_{ow}$ · Clmeta

   = 0.76 + 2.08

   = 2.84

   $\log K_{ow}$ of chlorpynfos = 1 + 2 + 3

   = 4.83

   experimental $\log K_{ow}$ = 4.81

Table 7. A variety of adjustments for groups in different types of chemical environments are described by Rekker,[72] who has also discussed inaccuracies and objections to this method of calculation.

c. *Fragmental Constants*

A major problem in using the substituent constants method described above is the difficulty of obtaining a log K_{ow} value for the parent compound. To overcome this problem a set of fragmental constants (f), representing structural fragments within complete molecules, has been developed, as reviewed by Rekker.[72] The log K_{ow} of a compound can be calculated from the following equation

$$\log K_{ow} = \sum_{n}^{n} a_n f_n$$

TABLE 8
Some Fragmental Constants Developed by Nye and Rekker

Aliphatic Compounds

Fragment	Constant	Fragment	Constant
H	0.21	OH	−1.440
CH	0.236	COO	−1.281
Cl	0.06	Br	0.24
C=O	−1.69	CH=CH$_2$	0.93
CH$_3$	0.702	CH$_2$	0.527
NH$_2$	−1.380	O	−1.536
C$_6$H$_5$	1.896	C (quat)	0.14

Aromatic Compounds

Fragment	Constant	Fragment	Constant
H	0.21	C$_6$H$_4$	1.719
Cl	0.943	COO	−0.43
OH	−0.359	NH	−0.93
C$_6$H$_5$	1.90	N	−1.06
COOH	0.00	C$_6$H$_3$	1.440

Proximity effects
  (pe1) Two hydrophilic groups separated by one CH$_2$: + .80
  (pe2) Two hydrophilic groups separated by two CH$_2$: + 0.46

Primary and secondary constants
  Those expressed to three decimal places are primary constants and those expressed to two
  decimal places are secondary constants.

Adapted from James, K. C., *Solubility and Related Properties*, Marcel Dekker, New York, 1986, 317 and 319.

where a is the numerical factor indicating the incidence of the fragment within the molecule, and the f the fragmental constant value.

Examples of fragmental constant values reported by Nyss and Rekker as recorded by James[75] are shown in Table 8. Thus, for ethyl acetate (H$_3$CCOOCH$_2$CH$_3$) the following calculation applies:

$$\log K_{OW} = 2f_{CH_3} + f_{CH_2} + f_{COO}$$

This calculation gives a log K_{OW} value of 0.65, which is comparable to the actual experimental value of 0.68.

The set of fragmental constants available has been expanded by substituting these known factors, referred to as primary values, into equations for compounds for which experimental values for log K_{OW} were available, thereby yielding a secondary set of f values as indicated in Table 8. In addition, corrections were introduced to allow for proximity effects of hydrophilic groups, which are also shown in Table 8. If hydrophilic groups are in close proximity, it reduces the number of water molecules which can associate with them as compared to when they are separated. This results in increased lipophilicity. Examples of some calculations based on these fragmental constants are set out in Table 9.

Another related approach based on fragmental constants was developed by Hansch and Leo.[76] They calculated that the fragmental constant for hydrogen (f$_H$) was 0.225. All other

TABLE 9

**Examples of Calculation of Log K_{OW} Using
Fragmental Constants**

1. 2-Aminoethanol

$$H_2N-CH_2-CH_2-OH$$

calc. log K_{OW} = $f_{NH_2} + 2 \times f_{CH_2} + f_{OH} + pe_2$

$$= -1.380 + (2 \times 0.527) - 1.440 + 0.46$$

calc. log K_{OW} = -1.31

exptl. log K_{OW} = 1.31

2. 4-Chlorophenol

calc. log K_{OW} = $f_{C_6H_4} + f_{Cl} + f_{OH}$

$$= 1.719 + 0.943 - 0.359$$

calc. log K_{OW} = 2.303

exptl. log K_{OW} = 2.42

values for fragmental constants were obtained by substituting this value into equations containing structural features for compounds for which reliable log K_{OW} values were available. In this manner, fragmental constants were calculated for single atom and multiple atom fragments, together with correction factors for the different structural situations in which these fragments were situated.

Lyman[47] has reported on this data and its application in log K_{OW} calculations. In tests on a range of different compounds he calculated that the error involved with this method was ± 0.12 log k_{OW} units. But he suggests that for more complex chemicals such as pesticides the error may be larger.

d. Parachor and Molar Volume

The mechanism involved in partitioning can be explained by the creation of a cavity in the solvent by molecules of the solute. The free energy involved in this change will be related to the size of the molar volume of the solute and the internal pressure of the solvent. James[77] has shown theoretically that for partitioning between water and organic solvents

$$\log K_{OW} = 0.2 \, V_2$$

where V_2 is the molar volume of the solute.

McGowan[78] used the "parachor" as a measure of molar volume and found the constant in the above equation to be 0.012 and that a correction, E_A, was needed to account for the interaction of solute and solvent molecules. Thus

$$\log K_{OW} = 0.012 \, [P] + E_A$$

where [P] is the parachor.

TABLE 10
Simplified Parachor Values

C – 9	S – 49
O – 20	F – 26
H – 15.5	Cl – 55
N – 17.5	Br – 68
P – 40	

double bond – 18
triple bond – 40

Note: Data from Briggs, G. G.

Lyman et al.[79] gives a comprehensive valuation of the various structural features which contribute to the overall parachor of a molecule. Briggs[6] has found the following relationship between parachor and log K_{ow}:

$$\log K_{ow} = 0.011\,P - 1.2\,n - 0.18 \qquad (r = 0.95)$$

where n is taken as 1 for each oxygen atom not bonded or conjugated to an aromatic ring (but including the ether oxygen in aryl alkyl ethers), one for each singly bonded nitrogen atom, one for each heterocyclic aromatic ring (however many heteroatoms it contains), and 0.25 for each halogen attached to a saturated carbon atom. In making his parachor calculations, Briggs used the parachor values shown in Table 10.

e. Molecular Connectivity

Molecular connectivity indices provide a method for calculating the molecular complexity of a molecule from its chemical structure. Indices in this family are also referred to as topological indices, and have been reviewed by Balaban et al.[80] These authors have referred to this family of indices as numerical quantities based on various invariants or characteristics of molecular graphs.

Probably the simplest index is the zero order connectivity index (^0X). It is the sum of the reciprocals of the square roots of the number of nonhydrogen atoms attached to each nonhydrogen atom in the molecule. Thus

$$^0\chi = \Sigma(\delta_i)^{-1/2}$$

where δ_i is the number of nonhydrogen atoms attached to each nonhydrogen atom. Thus $^0\chi$ = 4.121 for n-pentane. The first order connectivity index ($^1\chi$) is derived from each bond between two nonhydrogen atoms. The number of nonhydrogen atoms attached to the two atoms of each bond are counted, and the reciprocal of the square root of this is the bond value. The sum of each of the bond values is the $^1\chi$ value for the molecule. Thus, $^1\chi$ for n-pentane is 2.414. Similarly, an index can be obtained for the set of two adjacent bonds between nonhydrogen atoms in a molecule and another index obtained using three adjacent bonds. These indices are second and third order connectivity indices ($^2\chi$, $^3\chi$), respectively.

Murray et al.[81] have correlated the log K_{ow} values for sets of related compounds (alcohols, esters, ketones, etc.) with their first order connectivity indices yielding high correlation coefficients.

f. Collanders Equation

Partition coefficients (K_{Lw}) have been measured in a variety of water and immiscible organic solvent systems. This information provides a base from which the relationship

TABLE 11
Constants Derived for Some
Applications of Collanders Equation[a]

Solvent	x	y	r

Group A Solutes[b]

Solvent	x	y	r
Cyclohexane	1.481	2.729	0.761
Xylene	1.062	1.798	0.963
Oils	0.910	.1.192	0.981

Group B Solutes[c]

Solvent	x	y	r
Cyclohexane	0.941	0.690	0.957
Xylene	0.974	0.579	0.986
Oils	0.894	0.290	0.988

Solute Effect Not Significant

Solvent	x	y	r
Oleyl alcohol	1.001	0.576	0.985
Ethyl acetate	1.073	−0.056	0.969
Cyclohexanone	1.342	−1.162	0.985

[a] $\log K_{OW} = x \log K_{SW} + y$
[b] H-donors: acids, phenols, alcohols, etc.
[c] H-acceptors: hydrocarbons, ethers, esters, etc.

Adapted from Lyman, W. J., *Handbook of Chemical Property Estimation Methods*, McGraw-Hill, New York, p1.

between $\log K_{LW}$ and $\log K_{OW}$ values can be determined. Collanders equation describes a relationship which can be used to convert $\log K_{LW}$ values to $\log K_{OW}$ values and states

$$\log K_{OW} = a \log K_{LW} + b$$

where a and b are empirically developed constants.

Lyman et al.[7] and James[75] have reviewed the data available and the use of Collanders equation. Many systems and solutes have been investigated and systematized according to Collanders equation. Thus, a selection of values for the constants a and b are available according to the solute type and the solvent to water system in which its partition coefficient was measured. Examples of some of the information available are shown in Table 11.

g. Water Solubility

Hansch et al[82] investigated the relationship between water solubility, S, and the octanol-to-water partition coefficient for organic compounds and derived an equation of the form

$$\log (1/S) = a \log K_{OW} + b$$

where a and b are empirical constants for different groups of compounds. This relationship is valuable for calculating $\log K_{OW}$ values from S or vise versa. Miller et al.[83] have briefly reviewed recent work on this relationship.

Mackay et al.[84] and Chiou et al.[85] have provided a theoretical background to this relationship. Fugacity is a useful property for considering the relationship between solutes in different phases. It is defined as

$$f = C/V$$

where f is the fugacity of the compound, C the concentration, and V the fugacity capacity constant of the solute in the phase. Thus, fugacity is linearly related to concentration at the low concentrations which are typical of environmental xenobiotic chemicals. for a solute in solution its fugacity is

$$f = X_w Y_w f_R$$

where X_w is the solute mole fraction, v_w the aqueous activity coefficient on Raoults Law Basis, and f_R the reference fugacity or approximately the vapor pressure of the pure solute in a liquid form. For a pure liquid solute

$$f = f_R = X_w Y_w f_R$$

and

$$Y_w = 1/X_w$$

but

$$S = X_w M/V_w$$

where S is the solubility in mass to volume units, M the solute molecular weight, and V_w the molar volume of water. Thus

$$Y_w = M/SV_w$$

for liquid solutes.

Mackay et al.[84] have shown that for solids a correction is necessary and

$$Y_w = [M/SV_w][exp(6.791 - T_M/T]$$

where T_M is the solute melting point and T the system temperature.

These equations assume that the solute is at high dilution and does not affect the molar volumes.

When the octanol (designated by subscript O) and water phases are equilibrium, then

$$f = X_o Y_o f_R = X_w Y_w f_R$$

and

$$X_o Y_o = X_w Y_w$$

but

$$K_{ow} = (X_o V_w)/X_w V_o$$

thus

$$K_{ow} = Y_w V_w/Y_o V_o$$

and for liquids

$$K_{ow} = M/(SY_o V_o)$$

and for solids

$$K_{ow} = [exp(6.79(1 - T_M/T)M]/SY_o V_o$$

By taking natural logarithms of the above equations the following are obtained.

$$log \, K_{ow} = ln \, M - ln \, Y_o V_o - ln \, S \qquad \text{(for liquids)}$$

TABLE 12
Correlations of Octanol to Water Partition Coefficients
(Log K_{OW}) with Aqueous Solubility (Log S)

Type of compound	x	y	Number	r
Liquid organics				
Alcohols	−8.898	0.832	41	0.967
Ketones	−0.813	0.586	13	0.980
Esters	−0.987	0.513	18	0.990
Ethers	−0.846	0.791	12	0.938
Alkylhalides	−0.819	0.681	20	0.928
Alkynes	−0.773	0.806	7	0.953
Alkenes	−0.773	0.192	12	0.985
Aromatics	−1.004	0.340	16	0.975
Alkanes	−0.808	−0.201	16	0.953
All compounds	−0.747	0.730	156	0.935
Solid organics				
Organophosphates	−0.747	0.472	10	0.969
PCBs and DDT	−0.518	2.222	7	0.997
Ideal line	−1.000	0.92	—	—

Note: log K_{OW} = x log S + y; where S is in moles/liter.

From Chiou, C. T., Schmedding, D. W. and Manes, M., *Environ. Sci. Technol.*, 16, 4, 1982. Copyright American Chemical Society. With permission.

$$\log K_{OW} = \ln M + 6.79(1 - T_M/T) - \ln Y_O V_O - \ln S \quad \text{(for solids)}$$

These equations indicate an inverse correlation between aqueous solubility and octanol-to-water partition coefficient. The molar volumes for octanol and water (V_O, V_w) used do not take into account the solubilization of each of these solvents in the other, as outlined by Chiou et al.,[86] although the results of Miller et al.[83] indicate that this effect may be small.

A set of empirically determined relationships obtained by Chiou et al.[86] is shown in Table 12. These authors calculated the ideal line, defined by equations of a similar form to those derived above, as shown in Table 12. Some of the deviations of the observed relationships from the ideal line are attributed to inaccuracy in the basic data used. The solids show the greatest deviations from the ideal line, but when these are converted into the solubility of supercooled liquids by melting point correction, as shown above, the relationships are greatly improved.

B. WATER SOLUBILITY (S, S_w, C^s, C^s_w)
1. Definition and Nature

Interest in the relationship between water solubility and physiological properties began during the 1890s with Richet,[87] who found that with many compounds the toxic effects were inversely related to their solubility. Since that time the partition coefficient has become the principal physicochemical parameter used in quantitative structure activity relationships, but water solubility also has had extensive use.

A water solution is a homogeneous dispersion of individual molecules of a solute in water. The composition of an aqueous solution can be varied from zero concentration of solute up to a point where no more solute will dissolve, and a new phase, the undissolved solute, is evident. At this latter point the aqueous solution is saturated, and the concentration of solute in water is referred to as the water or aqueous solubility, referred to here as S when using mass to volume units, and S_w when using mole to volume units. It can be seen that aqueous solubility can be considered to be a partition process between the solute itself and water.

With organic compounds of low solubility in water, apparent solutions can be formed. The solute at concentrations significantly greater than the maximum water solubility forms small clusters of molecules, often called micelles or aggregates, which remain suspended in the water mass for long periods. In this situation observation indicates that solution of the solute has apparently occurred, since there is no evidence of another phase being formed. However, by filtering, allowing the suspension to stand for long periods, centrifuging and so on, such clusters can often, but not always, be removed and a true solution obtained. Thus, water solubility values reported in the literature can be higher than the true values, as discussed by Coates et al.[73]

Most environmental xenobiotic compounds which bioaccumulate are nonelectrolytes, such as chlorinated hycrocarbons and other hycrocarbons of various kinds. These substances have low overall polarity and generally do not contain polar functional groups. This results in a lack of ability to interact with water, and a corresponding inability to dissolve in this solvent. In addition, increasing molecular size results in increasing resistance by water to accept the molecule, and this also results in decreased water solubility. Some typical water solubilities for environmental xenobiotic compounds are shown in Table 1.

Expressions for the water solubility of a liquid solute with limited solubility in water have been developed by Chiou and Block,[88] and others. The molar solubility (S_w) is related to the activity coefficient (υ_w) by the following equation

$$S_w = 1/(Y_w V_w) \quad \text{(for liquids)}$$

where V_w is the molar volume of water.

For solids a correction is necessary due to the crystal interaction energy which is present in the solid, but not in the liquid. So to obtain the solubility of the supercooled liquid, which is dependent on the melting point of the solute, a correction is necessary as expressed in

$$S_w = [1/(Y_w V_w)][\exp 6.79(1 - T_M/T)]$$

where T_M is the solute melting point and T the system temperature.

This allows the calculation of solubilities, for the supercooled liquids corresponding with the solid solutes, which will be comparable to the solubility values for liquids. Yalkowsky et al.[93] have found the following expression also to be satisfactory for this correction purpose.

$$\log(S_w^C/S_w^L) = -0.01(T_M - 25)$$

where S_w^C and S_w^L are the solubility of the solute as a solid and a liquid, respectively.

Most measurements of water solubility of environmental xenobiotic compounds are expressed in units of mg L^{-1} or g m^{-3}, or equivalent units with μg L^{-1} being used for compounds having solubilities in the lower ranges. Occasionally, parts per million (ppm) and parts per billion (ppb) are used, which are usually equivalent to mg L^{-1} and μg L^{-1}, but could be mg kg^{-1} and g kg^{-1}, respectively.

It can be expected from the equations outlined above that with solids there will be a variation in solubility with temperature. In addition, there is usually a variation with tem-

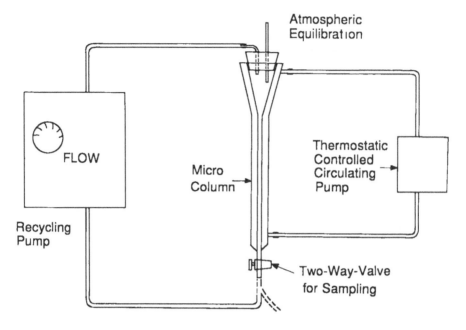

FIGURE 8.   Schematic illustration of the elution column test system for measuring water solubility. (From OECD Guidelines for Testing Chemicals, Organisation for Economic Co-operation and Development, Method 105, Paris, 1981. Copyright OECD. With permission.)

perature found with liquids. Water solubilities are usually measured at a standard temperature of 20°C, or in some cases, 25°C. The presence of electrolytes has been reported by James[90] to usually cause a "salting out" effect and a reduction in solubility, but in some cases an increase has been observed. In addition, the presence of nonelectrolytes in aqueous solution enhances the solubility of nonelectrolyte solutes depending on the concentration of the solvent nonelectrolyte, the similarity of the nonelectrolyte solute and solvent, and other factors. The degree of enhancement can be calculated and has been described by James.[90]

2. Experimental Measurement

James[89] has reviewed the wide variety of methods for measurement of water solubility in which the water and the solute are mechanically mixed, and the changes in concentration followed over time. When the aqueous solubility point is attained, this is evident from the plot of the results. Furer and Geiger[94] have described an optical method for determining the presence of fine suspensions and its application to solubility determinations. These methods often suffer from problems with the formation of molecular aggregates, as described previously. The time to achieve true equilibrium can be up to several months, as reported by Coates et al.,[72] and if not achieved, errors in the results can occur. Also, with many environmental xenobiotic compounds the solubility is at such a low concentration that the levels are difficult to quantitatively measure.

An improved method is described in the OECD Guidelines for Testing Chemicals[13] which uses an elution column, as illustrated in Figure 8. With this technique the solute is coated onto an inert support, and placed in a column which is externally thermostated to a constant temperature. Water is slowly pumped, or run under gravity, through the packed column and the eluent water concentration of solute measured. The concentrations are plotted over time until equilibrium is reached, and at this time the concentration of the now saturated solution is the water solubility.

Similarly, Wasik et al.,[55] May and Wasik,[91] Miller et al.,[56] and De Voe et al.[54] have used generator columns based on the same design as the elution column described above to determine the aqueous solubility of a variety of environmentally significant compounds. The methods used to process the eluent vary in both the collection method and the analytical technique, although gas chromatography and HPLC are the most commonly used analytical techniques. These methods reduce many of the difficulties of extended equilibrium times and molecular aggregate formation associated with the mechanical shaking or mixing methods described previously.

As mentioned initially, water solubility is in fact a partition process between the solute itself and water. This means that chromatographic behavior, as described in Section III.A.2. of this chapter for calculation of K_{OW}, should be applicable to water solubility. Thus, reversed phase chromatography could be applied with a nonpolar stationary phase resembling the nonpolar solute and a water solution for the mobile phase. Bruggeman et al.[69] have used reversed phase TLC with methanol in water solutions as eluent to investigate the solubility of aromatic hydrocarbons and PCBs. Later Swann et al.[65] used reversed phase HPLC with methanol in water mobile phase and obtained the following relationship with 15 diverse compounds including PCBs, DDT, and organophosphorus insecticides

$$\ln S(\text{mg L}^{-1}) = 7.618 \ln t_c - 0.01(T_M - 25) + 18.328 \qquad (r = 0.95)$$

where t_c is the corrected retention time of the solute in minutes.

Similarly, Coates et al.[73] used reversed phase HPLC to determine the water solubility of high molecular weight aliphatic hydrocarbons. A thorough evaluation of this technique to estimate solubility was carried out by Chin et al.[67] using reversed phase HPLC and methanol water as the mobile phase. Solubilities were estimated using the following equation

$$\log S = 1.74 - [\Delta S_f(R_M - T)]/2.3RT - \alpha \log t_c - \beta$$

where ΔS_f is the entropy of fusion of the solute, R the universal gas constant, and α and β empirically determined constants. The values of the constants α and β are determined by chromatographing compounds of known solubilities. For seventeen hydrocarbons and chlorinated hydrocarbons the relationship between the observed and calculated values for solubility was as follows:

$$\log S_{ob} = 0.980 \log S_p - 0.03 \qquad (r = 0.981)$$

where S_{ob} and S_p are the observed and predicted solubilities.

These authors suggest that this HPLC system is capable of giving reasonable accuracy in predicting the solubilities' appropriate unknown compounds.

3. Calculation of Water Solubility

A variety of methods for calculating the water solubility from a variety of other types of data are available as reviewed by Lyman[95] and also by James.[96] Table 13 summarizes the available techniques. The method of calculating solubility from log K_{ow} and vice versa, using regression equations from known sets of compounds, has been described in Section III.A.3.g. in this chapter, and so will not be rediscussed here. Lyman[95] has further listings of equations for this relationship with a more extended set of data than shown in Table 12. Some of the various methods available are briefly outlined below.

a. Extrapolation and Interpolation with Homologous Series

This method is analogous to the substituent constants method used with the calculation of K_{ow}, but is of more limited application. Coates et al.[73] found the following relationships with homologous series of hydrocarbons:

TABLE 13
Some Common Methods Used to Calculate Aqueous Solubility

Method	Information required	Comments
Extrapolation and interpolation with homologous series	Regression equation for relevant series	Accurate Applicable only to compounds in appropriate series
Regression equations with log K_{ow}	log K_{ow}	Fairly accurate Applicable only to compounds of known K_{ow}
Calculation from molar volume and parachor	Structure	Fairly accurate Fairly widely applicable
Calculation from cavity surface area	Structure	Fairly accurate Limited to sparingly soluble volatile liquids
Molecular connectivity	Structure	Fairly accurate Applicable only to compounds in appropriate series

n-alkenes	$\log S = 4.416 - 0.569\ N_c$
1-alkenes	$\log S = 5.108 - 0.569\ N_c$
2-methyl alkenes	$\log S = 4.559 - 0.569\ N_c$
3-methyl alkenes	$\log S = 4.376 - 0.559\ N_c$

In these series the methylene group shows a consistent contribution of approximately 0.57 per unit. This is comparable to 0.50 for the π constant used for methylene in the substituent constants values shown in Table 6. Also Nye and Rekker's fragmental constant of 0.527 for aliphatic compounds shown in Table 8 has a similar value.

James[96] has reviewed this application and reported that all hydrocarbon series together give less satisfactory relationships. However, the following equation, which is broadly applicable to alkanes, gives improved results

$$-\ln S_w = 1.49\ N_c - 0.321\ m + 0.787 \qquad (n = 14, \quad r = 0.998)$$

where S_w is in molar units and m represents the number of CH_3 groups.

b. Use of Molar Volume and Parachor

The basis of this method of calculation was outlined in relation to the calculation of log K_{ow} in Section III.A.3.d in this chapter. James[96] has described the calculation of parachor values, [P], according to the system in Table 14 and the calculation of solubility using

$$-\log S_w = -\log C_L + 0.0134\ [P]$$

where S_w is in mole kg,$^{-1}$ C_L the concentration of solute in pure liquid in mole kg.$^{-1}$ An example of calculations using this method is shown in Table 15.

Lande and Banerjee[98] have produced a set of equations for various series of compounds which relate water solubility to molar volume for organic nonelectrolytes. This includes many polyaromatic hydrocarbons (PAHs) and halogenated benzenes of environmental interest.

TABLE 14
Atomic Parachor Values
Abstracted from McGowan[a]

H – 24.7	O – 35.25
C – 46.35	S – 64.95
N – 40.8	Cl – 59.4

Note: Parachor calculated by addition
and subtraction of 18.6 for every
bond, whether single, double or
triple.

[a] Data from James.[97]

TABLE 15
Example of Calculation of Solubility Using the
Parachor

Benzene

$$P = (24.7 \times 6) + (46.35 \times 6) - (12 \times 18.6)$$
$$= 203.1$$
$$C_L = 1000/78 = 12.8 \text{ mole kg}^{-1}$$
$$-\log S_w = -\log 12.3 + (0.0134 \times 203.1)$$
$$= -1.615$$

thus

Calculated S_w = 0.024 mole kg^{-1}

Observed S_w = 0.018 mole kg^{-1}

c. Calculation from Cavity Surface Area

The solubility of nonelectrolytes in water can be related to the size of the cavity in the solvent created by molecules of solute.[92] A variety of methods have been used to estimate total cavity surface area (TSA), or the number of solvent molecules which can surround a solute molecule, from three dimensional physical models to computer based surface area calculations. In recent times computer based methods of calculation has become more readily available for calculating the TSA. Thus, it can be expected that this parameter will be more intensively investigated in the future. James[96] has reviewed a variety of the other methods used and their success in predicting solubility.

Lande and Banerjee[98] have investigated the application of TSA to many compounds of environmental interest. They found equations relating water solubility to TSA having the following form:

$$\log S_w = A_1 (TSA) + C_1 \text{ (for liquids)}$$

$$\log S_w = A_2 (TSA) + B_2(t) + C_2 \text{ (for solids)}$$

where A_1, A_2, B_2, C_2 are empirically developed constants.

The correlation coefficients within chemical classes were generally high ($r^2 > 0.93$) and also of about the same order as similar correlations with molar volume.

d. Calculation Using the Connectivity Index

Connectivity indices have been discussed in Section III.A.3.c. in this chapter. Hall et al.[99] have found that the first order connectivity index ($^1\chi$) to give the following relationships alcohols $\quad \ln S_w = 6.702 - 2.666\ ^1\chi \quad (r = 0.978, \quad n = 51)$

hydrocarbons $\quad \ln S_w = -1.505 - 2.533\ ^1\chi \quad (r = 0.958, \quad n = 18)$

These relationships have been developed using branched cyclic and straight chain compounds. This suggests that the technique may be applicable to more broadly based groups of compounds than some of the other techniques.

REFERENCES

1. **Woodwell, G. M.**, Toxic substances and ecological cycles, *Sci. Am.*, 216, 24, 1967.
2. **James, A. P. and Martin, A. J. P.**, Gas-liquid partition chromatography: The separation and micro-estimation of volatile fatty acids from formic acid to dodecanoic acid, *Biochem. J.*, 50, 679, 1952.
3. **Hamelink, J. L., Waybrant, R. C. and Ball, C.**, A proposal: exchange equilibria control the degree chlorinated hydrocarbons are biologically magnified in lentic environments, *Tras. Am. Fish. Soc.*, 100, 207, 1971.
4. **Hansch, C.**, A quantitative approach to biochemical structure-activity relationships, *Acc. Chem. Res.*, 2, 232, 1969.
5. **Connell, D. W.**, Bioaccumulation behaviour of persistent organic chemicals with aquatic organisms, *Rev. Environ. Contam. Toxicol.*, 101, 117, 1988.
6. **Briggs, G. G.**, Theoretical and experimental relationships between soil adsorption, octanol-water partition coefficients, water solubility, bioconcentration factors and the parachor, *J. Agric. Food Chem.* 29, 1050, 1981.
7. **Lyman, W. J., Reehl, W. F., and Rosenblatt, D. H.**, *Handbook of Chemical Property Estimation Methods*, McGraw-Hill, New York, 1982.
8. **Hawker, D. W. and Connell, D. W.**, Relationships between partition coefficient, uptake rate constant, clearance, and time to equilibrium for bioaccumulation, *Chemosphere*, 14, 1205, 1985.
9. **Branson, D. R., Blau, G. E., Alexander, H. C., and Neely, W. B.**, Bioconcentration of 2,2',4,4'-tetrachlorobiphenyl in rainbow trout as measured by an accelerated test, *Trans. Am. Fish. Soc.*, 104, 785, 1975.
10. **Hamelink, J. L.**, Current bioconcentration test methods and theory, in *Aquatic Toxicology and Hazard Evaluation*, ASTM STP, 634, Mayer, F. L. and Hamelink, J. L., Eds., American Society for Testing Materials, Philadelphia, 1977, 149.
11. **Oliver, B. G. and Niimi, A. J.**, Bioconcentration factors of some halogenated organics for rainbow trout: limitations in their use for prediction of environment residues, *Environ. Sci. Technol.*, 19, 842, 1985.
12. **Mackay, D.**, Correlation of bioconcentration factors, *Environ. Sci. Technol.*, 16, 274, 1982.
13. *OECD Guidelines for Testing of Chemicals*, Organisation for Economic Co-operation and Development, Paris, 1981, with periodic amendments.
14. **Garten, C. T. and Trabalka, J. R.**, Evaluation of models for predicting terrestrial food chain behaviour of xenobiotics, *Environ. Sci. Technol.*, 17, 590, 1983.
15. **Sugiura, K., Mase, T., Imasaki, K., and Goto, M.**, Accumulation and excretion of isopropylchlorobiphenyls in mouse and fish, *Bull. Environ. Contam. Toxicol.*, 26, 46, 1981.
16. **Geyer, H. J., Scheunert, I., and Korte, F.**, Distribution and bioconcentration potential of the environmental chemical pentachlorophenol (PCP) in different tissues of humans, *Chemosphere*, 16, 887, 1987.
17. **Canton, J. H., Wegman, R. C. C., Vulto, T. J. A., Verhoef, C. H., and Van Esch, G. J.**, Toxicity, accumulation and elimination studies of hexachlorocyclohexane (HCH) with salt water organisms of different trophic levels, *Water Res.*, 12, 687, 1978.
18. **Kenaga, E. E.**, Correlation of bioconcentration factors of chemicals in aquatic and terrestrial organisms with their physical and chemical properties, *Environ. Sci. Technol.*, 14, 553, 1980.
19. **Oliver, B. G. and Niimi, A. J.**, Bioconcentration of chlorobenzenes from water by rainbow trout: correlations with partition coefficients and environmental residues, *Environ. Sci. Technol.*, 17, 287, 1983.
20. **Davies, R. P. and Dobbs, A. J.**, The prediction of bioconcentration in fish, *Water Res.*, 18, 1253, 1984.
21 **Neely, W. B., Branson, D. R., and Blau, G. E.**, Partition coefficient to measure bioconcentration potential of organic chemicals in fish, *Environ. Sci. Technol.*, 13, 1113, 1974.
22. **Bishop, W. E. and Maki, A. W.**, A critical comparison of two bioconcentration test methods, in *Aquatic Toxicology* (ASTM STP 707), Eaton, J. G., Parrish, P. R., and Hendricks, A. C., Eds., American Society for Testing and Materials, 1980, 61.

23. **Sugiura, K., Ito, N., Matsumoto, N., and Mihara, Y.**, Accumulation of polychlorinated biphenyls and polybrominated biphenyls in fish: limitation of correlation between partition coefficients and accumulation factors, *Chemosphere*, 9, 731, 1978.

24. **Southworth, G. R., Keffer, C. C., and Beauchamp, J. J.**, Potential and realised bioconcentration. A comparison of observed and predicted bioconcentrations of azaarenes in the flathead minnow *(Pimephales pormelas)*, *Environ. Sci. Technol.*, 14, 1529, 1980.

25. **Muir, D. C. G., Yarechewski, A. L., and Knoll, A.**, Bioconcentration and disposition of 1,3,6,8-tetrachlorodibenzo-p-dioxin and octachlorodibenzo-p-dioxin by rainbow trout and fathead minnows, *Environ. Toxicol. Chem.*, 5, 261, 1986.

26. **Ellgehausen, H., Guth, J. A., and Esser, H. O.**, Factors determining the bioaccumulation potential of pesticides in the individual compartments of aquatic food chains, *Ecotoxicol. Environ. Saf.*, 4, 137, 1980.

27. **Geyer, H., Politzki, G., and Freitag, D.**, Prediction of ecotoxicological behaviour of chemicals: Relationship between n-octanol/water partition coefficient and bioaccumulation of organic chemicals by alga Chlorella, *Chemosphere*, 13, 262, 1984.

28. **Harding, G. C. and Vass, W. P.**, Uptake from seawater and clearance of DDT by marine planktonic crustacea, *J. Fish Res. Board Can.*, 37, 247, 1979.

29. **Crosby, D. G. and Tucker, R. K.**, Accumulation of DDT by *Daphnia magna*, *Environ. Sci. Technol.*, 5, 714, 1971.

30. **Southworth, G. R., Beauchamp, J. J., and Schmieder, P. K.**, Bioaccumulation potential of polycyclic aromatic hydrocarbons in *Daphnia pulex*, *Water Res.*, 12, 973, 1978.

31. **Sondergren, A.**, Uptake and accumulation of C^{14}-DDT by chlorella sp. (Chlorophyceae), *Oikos*, 19, 126, 1968.

32. **Baughman, G. L. and Paris, D. F.**, Microbial bioconcentration of organic pollutants from aquatic systems—a critical review, *CRC Crit. Rev. Microbiol.*, 8(3), 205, 1981.

33. **Vreeland, V.**, Uptake of chlorobiphenyls by oysters, *Environ. Pollut.*, 6, 135, 1974.

34. **Solbakken, J. E., Jeffrey, F. H., Knap, A. H., and Palmork, K. H.**, Accumulation and elimination of phenanthrene in the calico clam *(Macrocallista maculata)*, *Bull. Environ. Contam. Toxicol.*, 28, 530, 1982.

35. **Lee, R. F., Sauerheber, R., and Benson, A. A.**, Petroleum hydrocarbons: Uptake and discharge by the marine mussel *Mytilus edulis*, *Science*, 177, 344, 1972.

36. **Ernst, W.**, Determination of the bioconcentration potential of marine organisms — a steady state approach. I. Bioconcentration data for seven chlorinated pesticides in mussels *(Mytilus edulis)* and their relation to solubility data, *Chemosphere*, 11, 731, 1977.

37. **Renberg, L., Tarkpea, M., and Linden, E.**, The use of the bivalve *Mytilus edulis* as a test organism for bioconcentration studies. I. Designing a continuous-flow system and its application to some organochlorine compounds, *Ecotoxicol. Environ. Saf.*, 9, 171, 1985.

38. **Chadwick, G. G. and Brocksen, R. W.**, Accumulation of dieldrin by fish and selected fish-food organisms, *J. Wildl. Manage.*, 33, 693, 1969.

39. **Leatherland, J. F. and Sonstegard, R. A.**, Bioaccumulation of organochlorines by yearling coho salmon *(Oncorhynchus kisutch* Walbaum) fed diets containing great lakes Coho salmon and the pathophysiological responses of the recipients, *Comp. Biochem. Physiol.*, 72, C. 91, 1982.

40. **Reinert, R. E.**, Accumulation of dieldrin in the alga *(Scenedesumus obliquus)*, *Daphna magna*, and the guppy *(Poecilia reticulata)*, *J. Fish. Res. Board Can.*, 29, 1413, 1972.

41. **Metcalf, R. L., Sangha, G. K., and Kapoor, I. P.**, Model ecosystem for the evaluation of pesticide biodegradability and ecological magnification, *Environ. Sci. Technol.*, 5, 709, 1971.

42. **Bruggeman, W. A., Martron, L. B. J. M., Kooiman, D., and Hutzinger, O.**, Accumulation and elimination kinetics of di-tri- and tetrachlorobiphenyls by goldfish after dietary and aqueous exposure, *Chemosphere*, 10, 811, 1981.

43. **Geyer, H. J., Scheunert, I., Filser, J. G., and Korte, F.**, Bioconcentration potential (BCP) of 2,3,7,8-tetrachlorodibenzo-p-dioxin (2,3,7,8-TCDD) in terrestrial organisms including humans, *Chemosphere*, 15, 1495, 1986.

44. **Nernst, W.**, Vertheilung eines Stoffes zwischan zwei Losungsmitteln und zwischen Losungsmittel and Dampfraum, *Z. Phys. Chem.*, 8, 110, 1891.

45. **Leo, A., Hansch, C., and Elkins, D.**, Partition coefficients and their uses, *Chem. Rev.*, 71, 525, 1971.

46. **Hansch, C. and Dunn, W. J.**, Linear relationships between lipophilic character and biological activity of drugs, *J. Pharm. Sci.*, 61, 1, 1972.

47. **Lyman, W. J.**, Octanol/water partition coefficient, in *Handbook of Chemical Property Estimation Methods*, Lyman, W. J., Reehl, W. F., and Rosenblat, D. H., Eds., McGraw-Hill, New York, 1982, Chapter 1.

48. **Mackay, D.**, Solubility, partition coefficients, volatility and evaporation rates, in *Handbook of Environmental Chemistry*, Hutzinger, O., Ed., Springer-Verlag, Heidelberg, 1980, 31.

49. **Chiou, C. T.**, Partition coefficients of organic compounds in lipid-water systems and correlations with fish bioconcentration factors, *Environ. Sci. Technol.*, 19, 57, 1985.

50. **Hansch, C., Leo, A., Unger, S. H., Kim, K. H., Nikaitani, D., and Liem, E. J.**, Aromatic substituent constants for structure-activity correlations, *J. Med. Chem.*, 16, 1207, 1973.

51. **Fujita, T., Iwasa, J., and Hansch, C.**, A new substituent constant, II, derived from partition coefficients, *J. Am. Chem. Soc.*, 83, 5175, 1964.

52. **James, K. C.**, *Solubility and Related Properties*, Marcel Dekker, New York, 1986, 341.

53. **Tomlinson, E., Davis, S. S., Parr, G. D., James, M., Farraj, N., Kinkel, J. S. M., Glaisser, D., and Wynne, H. J.**, The filter probe extractor: a versatile tool for the rapid determination of solute oil/water distribution behaviour, in *Partition Coefficient, Determination and Estimation*, Dunn, W. J., Block, J. H., and Pearlman, R. S., Eds., Pergamon Press, Elmsford, NY, 1986, 83.

54. **DeVoe, H., Miller, M. M., and Wasik, S. P.**, Generator columns and high pressure liquid chromatography for determining aqueous solubilities and octanol-water partition coefficients of hydrophobic substances, *J. Res. Nat. Bur. Stand.*, 86, 361, 1981.

55. **Wasik, S. P., Miller, M. M., Tewari, Y. B., May, W. E., Sonnefeld, W. J., DeVoe, H., and Zoller, W. H.**, Determination of the vapor pressure, aqueous solubility and octanol/water partition coefficient of hydrophobic substances by coupled generator column/liquid chromatographic methods, *Residue Rev.*, 85, 29, 1983.

56. **Miller, M. M., Ghodbane, S., Wasik, S. P., Tewari, Y. B., and Martire, D. E.**, Aqueous solubilities, octanol/water partition coefficients and entropies of melting of chlorinated benzenes and biphenyls, *J. Chem. Eng. Data*, 29, 184, 1984.

57. **Woodburn, K. B., Doucette, W. J., and Andren, A. W.**, Generator column determination of octanol/water partition coefficients for selected polychlorinated biphenyl congeners, *Environ. Sci. Technol.*, 18, 457, 1984.

58. **Boyce, C. B. C. and Milborrow, B. V.**, A simple assessment of partition data for correlating structure and biological activity using thin layer chromatography, *Nature*, 208, 537, 1965.

59. **Tomlinson, E.**, Chromatographic hydrophobic parameters in correlation analysis of structure-activity relationships, *J. Chromatogr.*, 113, 1, 1975.

60. **Bowen, D. B., James, K. C., and Roberts, M.**, An investigation of the distribution coefficients of some androgen esters using paper chromatography, *J. Pharm. Pharmacol.*, 22, 518, 1970.

61. **Mirrlees, M. S., Moulton, S. J., Murphy, C. T., and Taylor, P. J.**, Direct measurement of octanol partition coefficient by high-pressure liquid chromatrography, *J. Med. Chem.*, 19, 615, 1976.

62. **Weber, W. J., Chin, Y. P., and Rice, C. P.**, Determination of partition coefficients and aqueous solubilities by reverse phase chromatography. I. Theory and Background, *Water Res.*, 20, 1433, 1986.

63. **McCall, J. M.**, Liquid-liquid partition coefficients by high-pressure liquid chromatography, *J. Med. Chem.*, 18, 549, 1975.

64. **Henry, B., Block, J. H., Anderson, J. L., and Carlson, G. R.**, Use of high-pressure liquid chromatography for quantitative structure-activity relationship studies of sulfonamides and barbiturates, *J. Med. Chem.*, 19, 619, 1976.

65. **Swann, R. L., Laskowski, B. A., McCall, J. M., Kuy, K. V., and Dishberger, H. J.**, A rapid method for the estimation of the environmental parameters octanol/water partition coefficient, soil sorption constant, water to air ratio and water solubility, *Residue Rev.*, 85, 17, 1983.

66. **Ellgehausen, H., D'Hondt, C., and Fuerer, R.**, Reverse-phase chromatography as a general method for determining octane-1-ol/water partition coefficients, *Pestic. Sci.*, 12, 219, 1981.

67. **Chin, Y. P., Weber, W. J., and Voice, T. C.**, Determination of partition coefficients and aqueous solubilities by reverse phase chromatography II. Evaluation of partitioning and solubility models, *Water Res.*, 20, 1443, 1986.

68. **Brooke, D. N., Dobbs, A. J., and Williams, N.**, Octanol: water partition coefficients (P): measurement, estimation, and interpretation, particularly for chemicals with $P < 10^5$, *Ecotoxicol. Environ. Saf.*, 11, 251, 1986.

69. **Bruggeman, W. A., Steen, J. V. D., and Hutzinger, O.**, Reversed-phase thin layer chromatography of polynuclear aromatic hydrocarbons and chlorinated biphenyls, relationship with hydrophobicity as measured by aqueous solubility and octanol-water partition coefficient, *J. Chromatogr.*, 238, 335, 1982.

70. **Rapport, R. A. and Eisenreich, S. J.**, Chromatrographic determination of octanol-water partition coefficients (K_{ow}s) for 58 polychlorinated biphenyl congeners, *Environ. Sci. Technol.*, 18, 163, 1984.

71. **Lyman, W. J.**, Estimation of physical properties, in *Environmental Exposure from Chemicals*, Volume I, Neely, W. B. and Blau, G. E. Eds., CRC Press, Boca Raton, 1985, 13.

72. **Rekker, R. F.**, *Hydrophobic Fragmental Constant*, Elsevier Scientific, Amsterdam, 1977.

73. **Coates, M., Connell, D. W. and Barron, D. M.**, Aqueous solubility and octan-1-ol to water partition coefficients of aliphatic hydrocarbons, *Environ. Sci. Technol.*, 19, 628, 1985.

74. **Neely, W. B.**, *Chemicals in the Environment: Distribution, Transport, Fate, Analysis*, Marcel Dekker, New York, 1980.

75. **James, K. C.**, *Solubility and Related Properties*, Marcel Dekker, New York, 317.

76. Hansch, C. and Leo, A., *Substituent Constants for Correlation Analysis in Chemistry and Biology*, John Wiley & Sons, New York, 1979.
77. James, K. C., *Solubility and Realted Properties*, Marcel Dekker, New York, 1986, 335.
78. McGowan, J. C., The physical toxicity of chemicals II. Factors Affecting Physical Toxicity in Aqueous Solutions, *J. Appl. Chem.*, 2, 323, 1952.
79. Lyman, W. J., Reehl, W. F. and Rosenblatt, D. H., *Handbook of Chemical Property Estimation Methods, Environmental Behaviour of Organic Compounds*, McGraw-Hill, New York, 1982, Chap. 12, 9.
80. Balaban, A. T., Motoc, I., Bonchev, D., and Mekenyan, O., Topological indices for structure-activity correlations, *Top. Curr. Chem.*, 114, 21, 1983.
81. Murray, W. J., Hall, L. H. and Kier, L. B., Molecular connectivity. III. Relationship to Partition Coefficients, *J. Pharm. Sci.*, 64, 1978, 1975.
82. Hansch, C., Quinlan, J. E., and Lawrence, G. L., The linear free energy relationship between partition coefficients and aqueous solubility of organic liquids, *J. Org. Chem.*, 33, 347, 1968.
83. Miller, M. M., Wasik, S. P., Huang, G. L., Shiu, W. Y., and McKay, D., Relationships between octanol-water partition coefficient and aqueous solubility, *Environ. Sci. Technol.*, 19, 522, 1985.
84. MacKay, D., Bobra, A., and Shiu, W. Y., Relationships between aqueous solubility and octanol-water partition coefficients, *Chemosphere*, 9, 701, 1980.
85. Chiou, C. T., Freed, V. H., Schmedding, D. W., and Kohnert, R. L., Partition coefficient and bioaccumulation of selected organic chemicals, *Environ. Sci. Technol.*, 11, 475, 1977.
86. Chiou, C. T., Schmedding, D. W., and Mains, M., Partitioning of organic compounds in octanol-water systems, *Environ. Sci. Technol.*, 16, 4, 1982.
87. Richet, M. C., Note sur le rapport entre la toxicitie et les properties physiques des corps, *Soc. Biol. Mem.*, 45, 775, 1893.
88. Chiou, C. T. and Block, J. H., Parameters affecting the partition coefficients of organic compounds in solvent-water and lipid-water systems, in *Partition Coefficient-Determination and Estimation*, Dunn, W. J., Block, J. H., and Pearlman, R. S., Eds., Pergamon Press, Elmsford, New York, 1986, 37.
89. James, K. C., *Solubility and Related Properties*, Marcel Dekker, New York, 36.
90. James, K. C., *Solubility and Related Properties*, Marcel Dekker, New York, 375.
91. May, W. E. and Wasik, S. P., Determination of the solubility behaviour of some polycyclic aromatic hydrocarbons in water, *Anal. Chem.*, 50, 997, 1978.
92. May, W. E., Wasik, S. P., and Freeman, D. H., Determination of the aqueous solubility of polynuclear aromatic hydrocarbons by a coupled liquid chromatographic technique, *Anal. Chem.*, 50, 175, 1978.
93. Yalkowsky, S. H., Valvani, S. C., and Mackay, D., Estimation of the aqueous solubility of some aromatic compounds, *Residue Rev.*, 85, 43, 1983.
94. Fürer, R. and Geiger, M., A simple method of determining the aqueous solubility of organic substances, *Pestic. Sci.*, 8, 337, 1977.
95. Lyman, W. J., Solubility in water, in *Handbook of Chemical Property Estimation Methods*, Lyman, W. J., Reehl, W. F. and Rosenblat, D. H., Eds., McGraw-Hill, New York, 1982, Chap. 2, 1.
96. James, K. C., *Solubility and Related Properties*, Marcel Dekker, New York, 1986, 355.
97. James, K. C., *Solubility and Related Properties*, Marcel Dekker, New York, 1986, 337.
98. Lande, S. S. and Banerjee, S., Predicting aqueous solubility of organic non-electrolytes from molar volume, *Chemosphere*, 10, 751, 1981.
99. Hall, L. H., Kier, L. B., and Murray, W. J., Molecular connectivity. II. Relationship to water solubility and boiling point, *J. Pharm. Sci.*, 64, 1974, 1975.

Chapter 3

GENERAL CHARACTERISTICS OF ORGANIC COMPOUNDS WHICH EXHIBIT BIOACCUMULATION

Des W. Connell

TABLE OF CONTENTS

I. INTRODUCTION

There is an enormous range of organic compounds from a wide variety of sources which have potential for distribution in the environment. Within this range it is helpful to identify the characteristics which distinguish the compounds exhibiting bioaccumulation from those which do not. However, the literature available does not allow an unequivocal identification of the specific characteristics required for bioaccumulation. In addition, the interaction of various factors and characteristics makes the isolation of specific factors which influence bioaccumulation difficult. Nevertheless, some of the general characteristics can be identified and discussed in broad terms as set out below.

Hansch[1] has suggested that biochemical quantitative structure-activity relationships (QSARs) would usually be expected to approximate a parabolic shape when a biological activity is related to a physicochemical property or structual characteristic for a set of compounds. Connell and Hawker[4] have shown that quantitative structure-activity relationships in bioaccumulation generally follow the same pattern. Thus, bioaccumulation would be expected to increase to reach a maximum, and then decline in relationship to increases in a specific characteristic or property of the set of chemicals being considered. Usually, the most important group of chemicals exhibiting most bioaccumulation fall on the rising arm of the parabola with a positive slope (see Figure 1). With many of these compounds there is an approximately linear relationship between bioaccumulation and the characteristic or property of the chemicals being considered in this region (see Figure 1). This general relationship should be kept in mind in the following discussion of the characteristics of compounds which exhibit bioaccumulation.

II. BASIC MOLECULAR CHARACTERISTICS

Molecular weight is a fundamental characteristic of a molecule which is easily obtained and bears a loose relationship to bioaccumulation. Brooke et al.[2] found that with 150 compounds of a wide variety of different types, significant bioaccumulation with fish started to occur at molecular weights of about 100 and increased with increasing molecular weight, reaching a maximum at molecular weight of about 350, and then declined with higher molecular weights. These features are illustrated graphically in Figure 1, in which the parabola is fitted to the chlorinated hydrocarbons alone, which appear to give the best relationship to molecular weight. Extrapolation of this parabolic relationship to the higher molecular weight compounds suggests that compounds with molecular weights of about 600 would have log K_B values of about zero. This is in accord with Zitko,[3] who suggested that compounds exhibiting bioaccumulation had an upper limit of molecular weight of 600 based on data on chloroparaffins, hexabromobenzene, and the PCBs. However, it can be seen from Figure 1 that the extrapolation of the parabola is based on a limited number of points, and that there are several compounds with molecular weights greater than 600 which exhibit some degree of bioaccumulation. Connell and Schüürmann[20] have noted that for chlorinated hydrocarbons and polyarmatic hydrocarbons there is a regular increase in bioaccumulation with fish, but the relationship is not highly significant.

Opperhuizen et al.[5] have pointed out that molecular weight is not a very reliable indicator of bioaccumulation behavior because octachloro benzo-p-dioxin (molecular weight 400) is not bioaccumulated while decachlorobiphenyl (molecular weight 499) shows an almost linear accumulation over time. To explain this, they suggest that the permeation concept of bioaccumulation requires "holes" in the hydrophilic outer surface of the membrane which separates the organism from water to be formed to allow entrance of the compound into the hydrophobic zone. Thus, the transport of a molecule is on the one hand dependent on the amount and size of the membrane "holes", and on the other, the molecular size of the

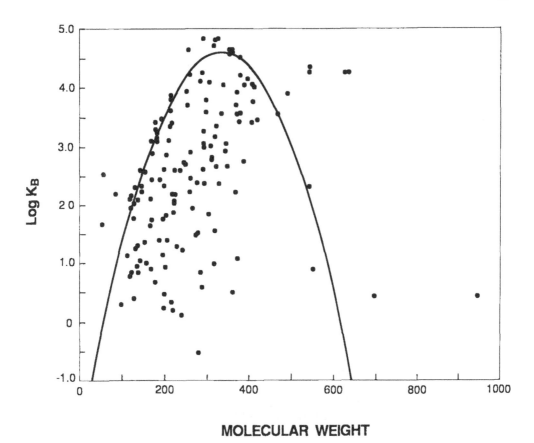

MOLECULAR WEIGHT

FIGURE 1. The relationship between log K_B and molecular weight for bioconcentration of 151 chemicals by fish. The parabolic curve is derived from the chlorinated hydrocarbons alone. (From Brooke D. N., Dobbs, A. J. and Williams, N., *Ecotoxicol. Environ. Saf.*, 11, 251, 1986. Copyright Academic Press. With permission.)

compound. Based on the data shown in Table 1, Opperhuizen et al.[5] have suggested that there is a loss of membrane permeation capability with hydrophobic molecules with widths over 9.5 Å. Thus, irrespective of molecular weight, compounds with cross sectional widths greater than 9.5 angstroms would not be expected to bioaccumulate. However, molecular widths generally would be expected to be reflected in an overall pattern by molecular weight.

Anliker et al.[6] have reported that various dye stuffs of relatively high molecular weight did not accumulate to any significant extent, although dye stuffs of lower molecular weight exhibited bioaccumulation which was predictable. Also, in later work Anliker and Moser[7] have reported that disperse dyes with molecular weights in the range of 360 to 546 exhibited limited bioaccumulation. They also indicated that all of these compounds had cross sectional areas greater than 9.5 Å and thus these results are basically in accord with the suggestion of Opperhuizen et al.[5] of a cross sectional area limit to bioaccumulation of 9.5 Å.

The value of the molecular weight as a characteristic of compounds which exhibit bioaccumulation probably reflects the importance of the size of the molecule in the bioaccumulation process. This aspect of bioaccumulation is discussed in Chapter 5—The Partition Mechanism. Connell and Schüürmann[20] found that the molecular surface area and the molecular volume reflected the bioaccumulation capacity of a group of chlorinated hydrocarbons and polyaromatic hydrocarbons. From their results it can be calculated that nonelectrolyte molecules with surface areas from 208 to 460 Å,[2] and molecular volumes from 260 to 760 Å,[3] are in the size ranges which usually exhibit bioaccumulation. The results of these authors

TABLE 1
Structural and Bioconcentration Parameters of Some Hydrophobic Chlorinated Aromatic Hydrocarbons

Name	M. Wt.	Cross section (Å)	log K_{ow}	log K_B
1,2-Dichloro naphthalene	197	8.6	4.9	3.4
1,4-Dichloro naphthalene	197	8.6	4.4	3.8
1,2,3,4-Tetrachloro naphthalene	266	9.1	5.9	4.5
1,3,5,7-Tetrachloro naphthalene	266	9.3	6.4	4.5
1,3,5,8-Tetrachloro naphthalene	266	9.3	6.0	4.4
Octachloro naphthalene	404	9.8	8.4	0
2,4,7,8-Tetrachloro-benzo-p-dioxin	322	7.6	>6	>3
Octachloro dibenzo-p-dioxin	460	9.8	>6	0
Hexachloro benzene	285	8.7	5.8	>4
Hexabromo benzene	552	9.6	>6	0
Decachloro biphenyl	499	8.7	>6	>5
2,2',3,3',4,4'-Hexabromobiphenyl	628	9.4	>6	>1
Decabromo biphenyl	944	9.6	>6	0

From Opperhuizen, A., Velde, E. W., Gobas, F. A. P. C., Liem, D. A. K., and Steen, J. M. D., *Chemosphere*, 14, 1891, 1985. Copyright Pergamon Press. With permission.

also suggest that compounds with areas and volumes above this range will bioaccumulate also, but to a lesser extent, and decreasing with increases in the parameter.

III. STRUCTURAL CHARACTERISTICS AND STABILITY OF COMPOUNDS

Within the molecular weight ranges discussed previously, a wide variety of compounds, representing many different structural types, exhibit bioaccumulation to some degree. Some examples of the structures of the compounds involved derived from MacKay[8] and Ellgehausen et al.[9] are shown in Figure 2. In addition, some examples of compounds which have very little bioaccumulation capacity, perhaps none which can be measured by current instrumental techniques, are also shown in Figure 2. These were derived from Opperhuizen et al.[5] and Anliker and Moser.[7] As a general rule, the chlorinated hydrocarbons and polyaromatic hydrocarbons exhibit highest capacity for bioaccumulation. A much more structurally diverse group of compounds have a more limited capacity for bioaccumulation, and some examples of these are shown in Figure 2.

The compounds exhibiting a high bioaccumulation capacity show a modest diversity of chemical structure, but most fall into the chlorinated and polyaromatic hydrocarbon groups previously mentioned. The parabolic curve in Figure 1 is based on the chlorinated hydrocarbons. Similarly, Connell and Hawker[4] found good parabolic relationships for the bioconcentration of chlorohydrocarbons by fish. A very limited range of types of chemical bonds are present in these groups. In most cases the bonds present are C–C aliphatic, C–C aromatic, C–H, C–Cl (or other halogen), but a limited representation of other bonds can be present in some compounds. These bonds are among the most stable which exist in organic compounds and confer exceptional stability on the associated compounds. This is reflected in the persistence of this group of compounds in soils, as indicated in Table 2.

Compounds with more limited capacity for bioaccumulation are shown in Figure 1 as points below the parabola. These compounds are structurally quite diverse, as indicated in Figure 2, and include a variety of organophosphorus compounds represented by profenofos. Thus, these compounds contain functional groups which are susceptible to attack by biotic or abiotic degradation processes. This leads to a limited persistence in the environment, and

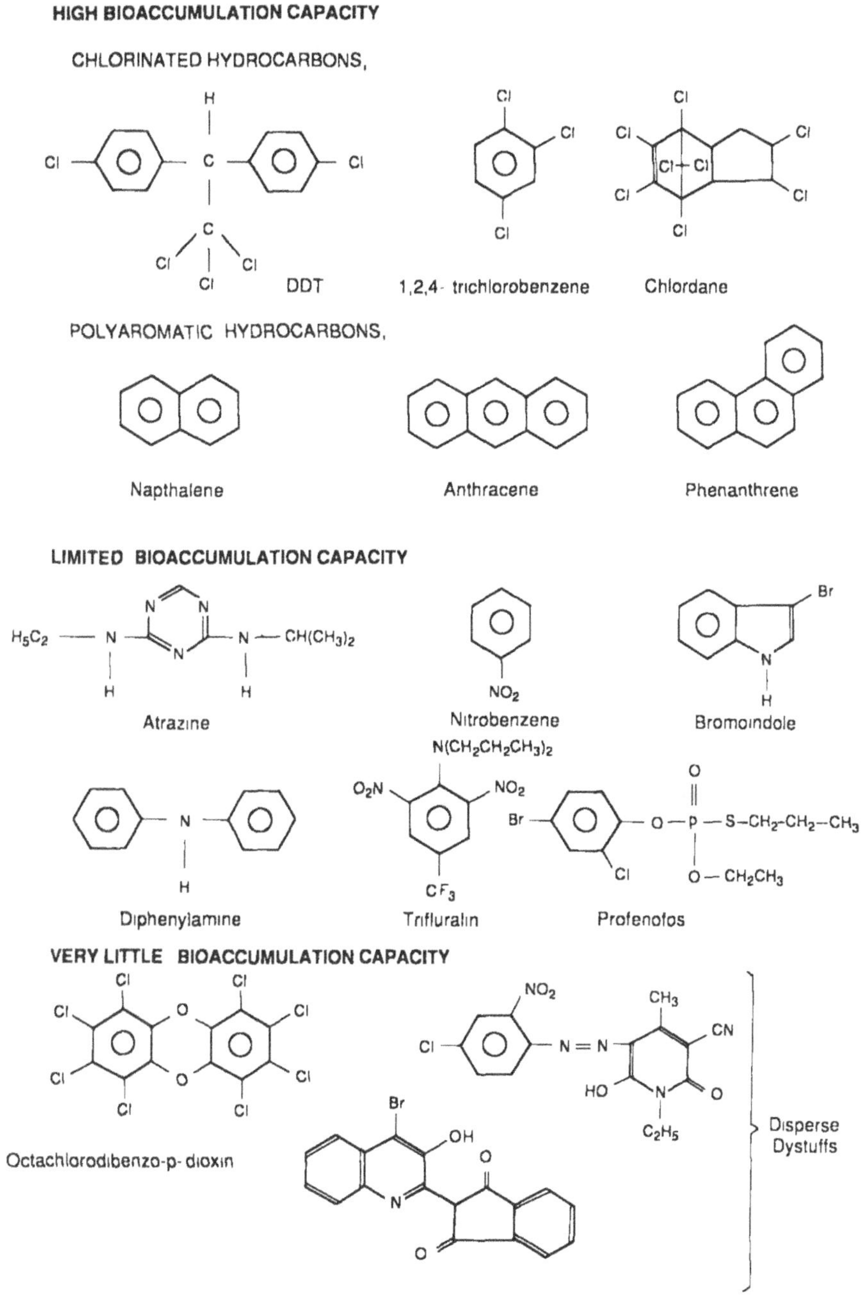

HIGH BIOACCUMULATION CAPACITY

CHLORINATED HYDROCARBONS,

DDT   1,2,4- trichlorobenzene   Chlordane

POLYAROMATIC HYDROCARBONS,

Napthalene   Anthracene   Phenanthrene

LIMITED BIOACCUMULATION CAPACITY

Atrazine   Nitrobenzene   Bromoindole

Diphenylamine   Trifluralin   Profenofos

VERY LITTLE BIOACCUMULATION CAPACITY

Octachlorodibenzo-p- dioxin

Disperse Dystuffs

FIGURE 2. Chemical structures of some examples of compounds with different bioaccumulation capacities.

within organisms after uptake has occurred, as illustrated in Table 2. On the other hand, these compounds generally contain lesser numbers of the very stable bonds which are present in the chlorinated and polyaromatic hydrocarbon groups.

The compounds which result in very little measurable bioaccumulation (see Figure 2) are also structurally very diverse. It is interesting to note here that the dye stuffs investigated by Anliker et al.[6] and Anliker and Moser[7] are very stable compounds. Examples of their structures are shown in Figure 2, and it can be seen that they contain many functional groups

TABLE 2
Persistence of Various Compounds in Soil

Group/compound	Persistence (years)[a]
Chlorinated hydrocarbons	2—5
Atrazine	0.8
Organophosphorus	0.02—0.23
Trifluralin	0.6

[a] Time taken for a pesticide applied to soil at the normal dosage to decrease by more than 75%.

Data from Connell, D. W. and Miller, G. J., *Chemistry and Ecotoxicology of Pollution*, John Wiley & Sons, New York, 1984.

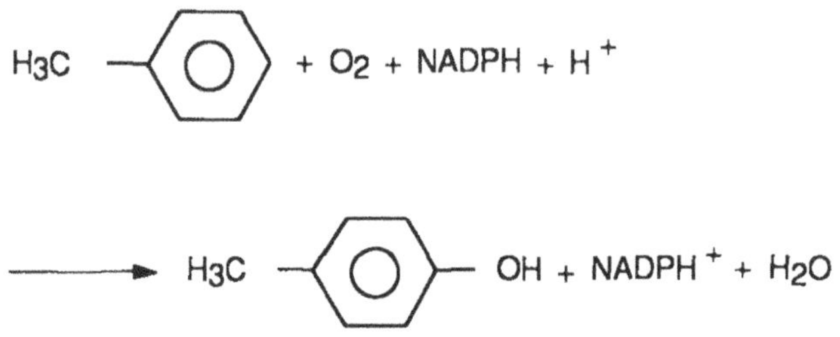

$$H_3C - \bigcirc + O_2 + NADPH + H^+$$

$$\longrightarrow H_3C - \bigcirc - OH + NADPH^+ + H_2O$$

FIGURE 3.   An example of the chemical transformations resulting from MFO activity in organisms.

which would be expected to be susceptible to chemical and biological attack. Possibly, the exceptional stability of these compounds in biota is due to their low water solubility, which provides a protection by making the compound unavailable for attack by these processes.

Most biota have a limited capacity to metabolize and excrete xenobiotic chemicals which bioaccumulate, due to these compounds containing structures which are resistant to chemical attack. The organism response on exposure to such xenobiotic compounds is the induction of formation of an oxidative enzyme system. This allows detoxification to proceed by an initial oxidation step, which results in the conversion of the highly lipophilic xenobiotic compounds into more hydrophilic oxidation products which the organism can process successfully. This oxidative metabolism is catalyzed by the hepatic microsomal enzyme system which contains cytochrome P-450 dependent mixed function oxidase (MFO) systems. An example of the chemical transformations which occur during this process is shown in Figure 3.

An important aspect of degradation and the stability of compounds in biota is the capacity of the biota to undergo enzyme induction, as outlined above. Moriarty and Walker[19] have pointed out that terrestrial animals usually have a higher metabolic capacity for lipophilic compounds than aquatic animals. This means that compounds will differ in their bioaccumulation capacity between aquatic and terrestrial animals, exhibiting generally lower bioaccumulation in terrestrial animals.

As well as overall chemical structure and bond types having an influence on bioaccumulation, there are more subtle structural aspects which can be important. With the PCBs, various authors[10,11,12] have found that the conformation of the PCB molecule has an influence on the amount of bioaccumulation which occurs. The conformation of PCB congeners is influenced by the chlorine substitution pattern. In particular the chlorines in the 2,6, and

$2'$-$6'$ in the molecule lead to a conformation of the molecule in which the two aromatic rings are nonplanar, and at an angle of somewhere around 90° to one another, in some cases. Shaw and Connell[10] have developed a set of correction coefficients to account for the effect of these stereochemistry factors on bioaccumulation. A substantial decline in bioaccumulation was observed for compounds with chlorines in all the 2,6, $2',6'$ positions.

IV. LIPOPHILICITY

As mentioned previously, compounds with a high bioaccumulation capacity usually contain a limited range of bond types (see Figure 2). These bonds are usually C–C (aliphatic), C–C (aromatic), C–H, C–Cl (or other halogens), and as well as being very stable, they are relatively nonpolar and give rise to nonpolar compounds. Nonpolar compounds would be expected to have low water solubility and high fat, or lipid, solubility.

Highly fat soluble, lipophilic compounds are the organic compounds most likely to bioaccumulate in biota. Mackay[8] has theoretically demonstrated that the lipid phase in biota is the dominant phase for the accumulation of these substances. Bioaccumulation of organic compounds increases with increasing lipophilicity, which is conveniently measured by the 1-octanol to water partition coefficient, K_{OW}. The octanol-to-water partition coefficient is the concentration in octanol/concentration in water for the compound distributed between the two phases at equilibrium. For compounds which have bioaccumulation potential, the K_{OW} usually ranges between about 100 and in excess of 10 million. Since these numbers are over an inconvenient range, K_{OW} is usually expressed as a log K_{OW} giving a range of values from 2 to about 7 for most of the compounds of interest. Connell and Hawker[4] and Connell[13] have found that bioaccumulation declines for compounds having log K_{OW} values above about 6, and Zitko[3] reports that it is difficult to determine for those with values below 2.

Mackay[8] has noted that water solubility is related to K_{OW} by equations of the general form, K_{OW} = a constant/water solubility; so bioaccumulation increases as solubility declines. In accord with this situation, bioaccumulation declines when solubilities are less than 0.02 to 0.002 moles/m^3. This is in general agreement with Zitko,[3] who has suggested that bioaccumulation declines with solubilities less than 0.005 moles/m^3.

A limited number of results are available which indicate an upper limit to log K_{OW} above which bioaccumulation will not occur. For example, Muir et al.[14] have conducted investigations into the bioaccumulation of compounds in the log K_{OW} range 6 to 10 with fish. These results suggest that there is an upper limit to bioaccumulation with compounds having log K_{OW} values of about 10 to 12. Anliker and Moser,[7] in their investigations of dyes, generally found that many dyes exhibited undetectable bioaccumulation at log K_{OW} values somewhat less than 10 to 12. This may be explained by the relatively high molecular weight of these substances and their having a molecular width in excess of 9.5 Å. More recently, Connell and Hawker[4] have reported a parabolic relationship between log K_{OW} and bioaccumulation with a maximum at log K_{OW} of 6.7. One of the reasons suggested by Connell and Hawker as well as Dobbs and Williams[15] for the declining bioaccumulation of these high molecular weight compounds was related to their reduced solubility in biota fat.

V. IONIZATION CHARACTERISTICS

The lipophilicity of nonpolar compounds can be substantially reduced by the introduction of a polar or ionizable group, such as an acid, phosphate or sulfonate group, into the molecule. For example, Esser and Moser[16] found that the log K_{ow} of naphthalene was reduced from about 3.5 to -1.5 by the introduction of a sulfonate group. It would be expected that this change in lipophilicity would be reflected in the relative bioaccumulation capacity of these two compounds.

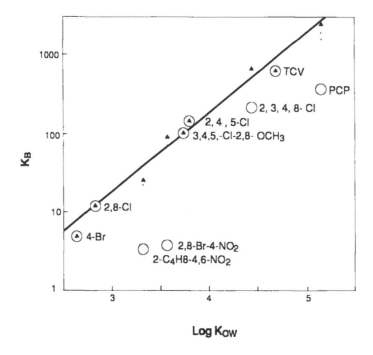

FIGURE 4.   Relation of the bioconcentration factor ($K_B = C_{fish}/C_{water}$) of eight phenols and tetrachloroveratrol at pH 6 to their octanol/water partition coefficients (K_{ow}). O, measured values: Δ K_B values corrected for ionization. The regression line was calculated for the corrected values. (From Saarikoski, J. and Viluksela, M., *Ecotoxicol. Environ. Saf.*, **6**, 501, 1983. Copyright Academic Press. With permission.)

This effect is well illustrated by the results of Saarikoski and Viluksela,[17] who have measured the bioaccumulation factor for a series of substituted pheols with fish. The results obtained are shown in Figure 4, and indicate that the highly ionized compounds do not follow the trend of the other relatively unionized substances. However, when all substances are corrected for ionization a good correlation is obtained.

Esser and Moser[16] point out several other effects of chemical conditions on log K_{ow} values. The presence of sodium chloride may result in the production of a "common ion" effect on solutions of sodium salts of neutral compounds resulting in the equilibrium between ionic and neutral forms being displaced towards the neutral compound. This would result in changing the proportions of the compound which would dissolve in octanol and in water, and thus a change in the log K_{ow} value would result (see Figure 5). Also, the pH of solutions of weakly ionized acids has an effect on the proportion of unionized compound which is present in solution, and therefore, the log K_{ow} value, as illustrated in Figure 6. As a result of these observations, Esser and Moser[16] suggest that for better standardization in the measurement of the octanol-to-water partition coefficient of compounds of the type mentioned above, the log K_{ow} value should be measured using a buffer at physiological pH and sodium chloride concentration, as the aqueous phase.

VI. OVERALL CHARACTERISTICS OF ORGANIC COMPOUNDS WHICH BIOACCUMULATE

It is clear that the general characteristics of compounds controlling bioaccumulation are interrelated. The nature of the individual bonds confers on the derived molecules the prop-

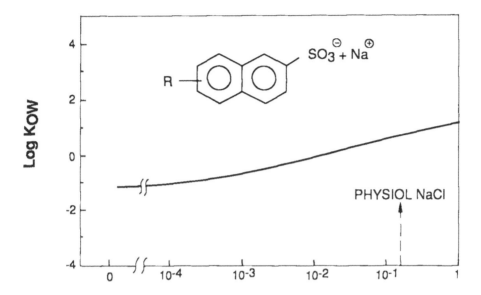

FIGURE 5.   Effect of neutral salt on the K_{ow} values of charged CPD due to ion pair extraction or the "common ion" effect. (From Esser, H. O. and Moser, P., *Ecotoxicol. Environ. Saf.*, 6, 13, 1982. Copyright Academic Press. With permission.)

erties of nonpolarity and resistance to degradation. Increasing molecular size results in an increasing partition coefficient and a consequent related decline in water solubility, resulting in an increase in the bioaccumulation capacity. However, at certain molecular sizes, molecular surface areas, and molecular volumes, as well as the related log K_{ow} and water solubility values, a decline in bioaccumulation occurs, possibly related to such factors as the reduced fat solubility of these higher molecular weight compounds and the lack of ability to penetrate biological membranes. These factors are set out in an abbreviated form in Table 3. It is noteworthy that these are general characteristics which are appropriate for the usual range of compound types, exposure periods, and kinetic behavior encountered in laboratory investigations. However, Connell[13] has reported that the bioaccumulation process takes a significant period of time to occur, which increases with increasing molecular weight and other related characteristics with most compounds. As a result, relatively long periods of time are required for compounds with comparative high partition coefficients. Thus, it is possible that in many cases with higher molecular weight compounds, equilibrium has not been attained, and thus the bioaccumulation reported is less than would be expected. This means that kinetic and other factors need to be taken into account to gain a clearer understanding of the behavior of chemicals which bioaccumulate. Also, with biota in actual aquatic ecosystems, bioaccumulation behavior may differ from that obtained in laboratory experiments due to different uptake routes and exposure times, etc.

Lech and Bond[18] have reviewed the mechanisms by which lipophilic compounds can be biologically degraded. Rapid degradation will not allow bioaccumulation to occur at all, but a moderate rate will result in bioaccumulation to a limited extent. However, sometimes different rates may occur in different groups of organisms or species. Thus, it could be expected that different biological groups may exhibit different bioaccumulation factors with the same compound.

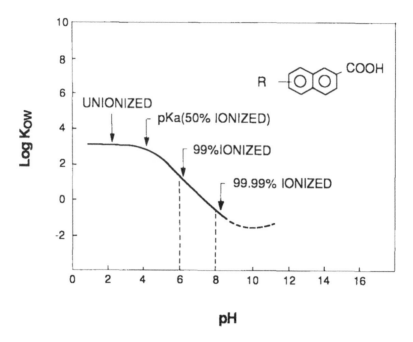

FIGURE 6.   Effect of pH on log K_{ow} of a weak acid. (From Esser, H. O. and Moser, P., *Ecotoxicol. Environ. Saf.*, 6, 13, 1982. Copyright Academic Press. With permission.)

TABLE 3
General Characteristics of Organic Chemicals Which Exhibit Bioaccumulation

Characteristic	Features giving bioaccumulation
Chemical structure	*High capacity:* high proportion of C–C (aliphatic), C–C (aromatic), C–H and C–halogen bonds
	Limited capacity: low proportion of the bonds above with the presence of variety of functional groups
Molecular weight	>100 giving a maximum capacity at about 350, then declining to very low capacity about 600
Molecular dimensions	Cross section width <9.5 Å molecular surface area between 208 and 460 Å² molecular volume between 260 and 760 Å³
Stability	Resistant to degradation reflected in soil persistence in the order of years
Log K_{ow}	>2 giving a maximum capacity of about 6 and a decline to very low capacity at about 10—12
Water solubility (mole m⁻³)	<18 giving a maximum at about 0.002 with declining capacity of lower values
Degree of ionization	Very low

REFERENCES

1. **Hansch, C.,** A quantitative approach to biochemical structure-activity relationships, *Acc. Chem. Res.*, 2, 232, 1969.
2. **Brooke, D. N., Dobbs, A. J., and Williams, N.,** Octanol: water partition coefficients (P): measurement estimation and interpretation, particularly for chemicals with P > 10⁵., *Ecotoxicol. Eviron. Saf.*, 11, 251, 1986.

3. **Zitko, V.**, Metabolism and distribution by aquatic animals, in *Handbook of Environmental Chemistry*, Hutzinger, O., Ed., Springer-Verlag, Berlin, 221, 1980.

4. **Connell, D. W. and Hawker, D. W.**, Use of polynomial expressions to describe the bioconcentration of hydrophobic chemicals by fish, *Ecotoxicol. Environ. Saf.*, 16, 242, 1988.

5. **Opperhuizen, A., Velde, E. W., Gobas, F. A. P. C., Liem, D. A. K., Steen, J. M. D., and Hutzinger, O.**, Relationship between bioconcentration in fish and steric factors of hydrophobic chemicals, *Chemosphere*, 14, 1871, 1985.

6. **Anliker, R., Clarke, E. A., and Moser, P.**, Use of the partition coefficient as an indicator of bioaccumulation tendency of dye stuffs in fish, *Chemosphere*, 10, 263, 1981.

7. **Anliker, R. and Moser, P.**, The limits of bioaccumulation of organic pigments in fish: their relation to the partition coefficient and the solubility in water and octanol, *Ecotoxicol. Environ. Saf.*, 13, 43, 1987.

8. **Mackay, D.**, Correlation of bioconcentration factors, *Environ. Sci. Technol.*, 16, 274, 1982.

9. **Ellgehausen, H., Guth, J. A., and Esser, H. O.**, Factors determining the bioaccumulation potential of pesticides in the individual compartments of aquatic food chains, *Ecotoxicol. Environ. Saf.*, 4, 134, 1980.

10. **Shaw, G. R. and Connell, D. W.**, Physicochemical properties controlling polychlorinated biphenyl (PCB) concentration in aquatic organisms, *Environ. Sci. Technol.*, 18, 18, 1984.

11. **Sugiura, K., Ito, N., Matsumoto, N., Nikora, Y., Murata, K., Tsukakoshi, Y., and Goto, M.**, Accumulation of polychlorinated biphenyls and polybrominated biphenyls in fish: limitation of correlation between partition coefficients and accumulation factors, *Chemosphere*, 9, 731, 1978.

12. **Hutzinger, O., Tulp, M. T. T. M., and Zitko, V.**, Chemicals pollution potential, in *Aquatic Pollutants—Transformations and Biological Effects*, Hutzinger, O., Lelyveld, L. V. H., and Zoeteman, B. C. J., Eds., Pergamon Press, Oxford, 1978, 13.

13. **Connell, D. W.**, Bioaccumulation behavior of persistent organic chemicals with aquatic organisms, *Rev. Environ. Contam. Toxicol.*, 101, 117, 1988.

14. **Muir, D. C. G., Marshall, W. R., and Webster, D. R. B.**, Bioconcentration of PCDDs by fish: effects of molecular structure and water chemistry, *Chemosphere*, 14, 829, 1985.

15. **Dobbs, A. J. and Williams, N.**, Fat solubility—a property of environmental relevance? *Chemosphere*, 12, 97, 1983.

16. **Esser, H. O. and Moser, P.**, An appraisal of problems related to the measurement and evaluation of bioaccumulation, *Ecotoxicol. Environ. Saf.*, 6, 131, 1982.

17. **Saarikoski, J. and Viluksela, M.**, Relations between physicochemical properties of phenols and their toxicity and accumulation in fish, *Ecotoxicol. Environ. Saf.*, 6, 501, 1983.

18. **Lech, J. J. and Bond, J. R.**, Relationship between biotransformation and the toxicity and fate xenobiotic chemicals in fish, *Environ. Health Perspect.*, 34, 115, 1980.

19. **Moriarty, F. and Walker, C. H.**, Bioaccumulation in food chains—a rational approach, *Ecotoxicol. Environ. Saf.*, 13, 208, 1987.

20. **Connell, D. W. and Schüürmann, G.**, Evaluation of various molecular parameters as predictors of bioconcent ration in fish, *Ecotoxicol. Environ. Saf.*, 15, 324, 1988.

Chapter 4

ENVIRONMENTAL ROUTES LEADING TO THE BIOACCUMULATION OF LIPOPHILIC CHEMICALS

Des W. Connell

TABLE OF CONTENTS

I. DISTRIBUTION PATTERNS OF LIPOPHILIC CHEMICALS IN THE ENVIRONMENT

The environment is highly complex, but can be considered to consist of a number of separate phases or compartments. These phases or compartments are segments of the environment within which a chemical can be considered to behave in a uniform manner. When a chemical is discharged to the environment, its distribution between these phases is governed by the physicochemical properties of the chemical and the properties of the phases within the environment. The phases most commonly used are air, soil, water, sediments, and biota; Figure 1 in Chapter 1 illustrates the distribution process.

Some of these phases are clearly more complex than others, and a refinement of the model can be obtained by subdividing the more complex phases into additional phases. For example, sediments can be subdivided into bottom and suspended sediments (see Chapter 1, Figure 2). Of course, for this to be successful, data must be available on the new phases and the behavior of the chemical in them. Clearly, this approach involves some oversimplification, but the more accurately the phases reflect the actual environment, the more precise the results obtained can be expected to be.

The property which governs the distribution of a chemical between phases is described as the "fugacity"[1]. In more descriptive terms this property is often referred to as the "escaping tendency". If sufficient data is available it can be accurately calculated, but has different values in different phases with the same compound. The fugacity in a particular phase is usually linearly related to its concentration at the low concentrations which are usually encountered with environmental contaminants. Thus

$$f = C/Z$$

where f is the fugacity, C the concentration, and Z the fugacity capacity constant. This means that the fugacity capacity constants must be known to calculate fugacity. These can be derived from data on physicochemical properties such as the bioaccumulation factor, Henry's Law Constant, the soil to water distribution coefficient, and so on.

The particular value of fugacity is that at equilibrium the fugacities in all the different phases are equal so

$$f_{equil} = f_{air} = f_{water} = f_{sediments} = f_{biota}$$

Thus, there is one value for fugacity (f_{equil}) which applies to that compound as it is distributed in each phase. It can be shown that

$$f_{equil} = M_t/\Sigma(Z_i V_i)$$

where M_t is the total mass of the substance discharged to the environment, Z_i the fugacity constant for each compound in each individual phase, and V_i the volume of each individual phase. So to apply this model, the volumes of the phases are needed, as well as the fugacity capacity constants in those phases. Once f_{equil} is known, this value can be applied to each phase, allowing the distribution and final concentrations in all the phases to be calculated.

Mackay and Patterson[1] have estimated the distribution of a variety of substances in a typical environment, as shown in Table 1. These calculations reflect an equilibrium situation in a static environment. But in fact, the atmosphere, water, and so on exhibit varying degrees of change and movement, and thus are dynamic. These dynamic factors influence the final concentrations which occur in the actual environment. Also, the final concentrations are affected by resistance to the transfer of chemical between phases. Mackay and Patterson[2]

TABLE 1
Calculated Percentage Mass Distributions for Selected Substances

	Naphthalene	Butyl benzyl phthalate	Tetrachloro-PCB	DDT
Air	91.5	0.71	94.0	2.35
Soil	0.35	15.5	0.98	17.1
Water	6.49	11.4	0.42	0.77
Biota	0.0005	0.013	0.00075	0.0093
Suspended solids	0.003	0.12	0.0076	0.13
Sediment	1.64	72.3	4.56	79.7

Compiled from Mackay, D. and Patterson, S., *Environ. Sci. Technol.*, 15, 1006, 1981.

have described ways in which the model can be expanded to account for these factors, and so more accurately reflect the actual environment.

This model includes the bioaccumulation of compounds from water by aquatic biota only, which is characterized by the bioaccumulation factor K_B. It demonstrates how the final levels of bioaccumulation are affected by the other equilibrium processes which exist in the natural environment. But in fact, the relationship used, in this case, has been established with fish only. Nevertheless, the information provided is valuable in that it gives an indication of likely levels of lipophilic contaminants in these organisms. However, bioaccumulation occurs with other aquatic organisms in addition to terrestrial organisms. Further calculations would be required to establish the likely levels of bioaccumulation with these organisms. But in many cases the factors influencing bioaccumulation in a range of organisms have not been sufficiently clearly established to allow incorporation into a model of this type.

II. PATHWAYS TO BIOTA

A. GENERAL ASPECTS OF THE MOVEMENT OF CHEMICALS TO BIOTA

The pathways of xenobiotic chemicals to biota, covered simply by the water-to-fish relationship only, as described previously in the fugacity based model, can be expanded diagrammatically to cover a wide range of organisms, as shown in Figure 1. In this representation, aquatic biota are those organisms which take up xenobiotic chemicals from water, and a water-to-organism equilibrium established. This includes fish, molluscs, many invertebrates, and so on. Terrestrial biota take up xenobiotic ehcmicals from soil, the atmosphere, and food, and do not establish an equilibrium with an ambient water mass; they include such organisms as aquatic birds and mammals. The relationships between concentrations in all pairs of phases at equilibrium have not been clearly characterized at the present time; for example, the concentration in food to concentration in predator relationship, as shown in Figure 1, with the low trophic level aquatic biota to higher trophic level biota, as well as the terrestrial vegetation to low trophic level terrestrial biota, and the low trophic level terrestrial biota to higher trophic level terrestrial biota. These relationships are not characterized by a directly analogous partition process to bioconcentration by fish, although the evidence suggests that partitioning is involved. This is discussed in detail in Chapter 7, Biomagnification of Lipophilic Compounds in Terrestrial Systems. Since a partition coefficient cannot be assigned to these processes at present, they are distinguished by a single headed arrow in Figure 1.

On the other hand, the current evidence indicates that there is an equilibrium between the remaining pairs of phases, which can be characterized by a partition coefficient. This has been well established in some cases, such as the water-to-fish equilibrium, represented by the water-to-aquatic biota equilibrium in Figure 1. In other cases, such as terrestrial biota-to-air process, the data available is limited. Recent information has shown that some infauna,

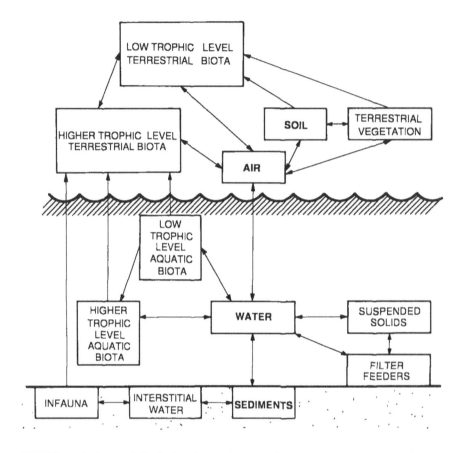

FIGURE 1.   Pathways of distribution of a persistent xenobiotic chemical among the biotic and abiotic components of the natural environment.

in particular polychaetes, probably acquire lipophilic compounds in the sediments from the interstitial water, and that the process is controlled by a partition mechanism.

The fugacity modeling previously described (see Table 1), and the pathways of distribution of a chemical in the environmental phases (see Figure 1), indicate that for most lipophilic compounds two major abiotic systems exist in the natural environment which regulate the exposure of biota to chemicals. These are the sediment-to-water system and the soil-to-atmosphere system which contain the major proportions of lipophilic chemicals after distribution occurs, and provide the routes to almost all biota. The soil and sediments act as a reservoir releasing compounds to the atmosphere and water resulting in the exposure of most biota. Also, the atmosphere-to-vegetation route could be an important exposure route for many terrestrial biota.

B. THE SEDIMENT-TO-WATER SYSTEM

Aquatic areas are an important sink for many lipophilic chemicals discharged to the environment due to a variety of factors. In some situations, aquatic areas are a substantial segment of the environment, for example, those areas which include large lakes, rivers or coastal areas. However, in some other environments such as deserts and arid regions, aquatic areas are not present. But suspended matter in runoff water during the occasional wet periods would be expected to be deposited in lower zones, resulting in an accumulation of xenobiotic chemicals in these areas. Many lipophilic compounds, such as agricultural chemicals which originate from terrestrial sources, enter aquatic areas as surface sorbed materials on eroded

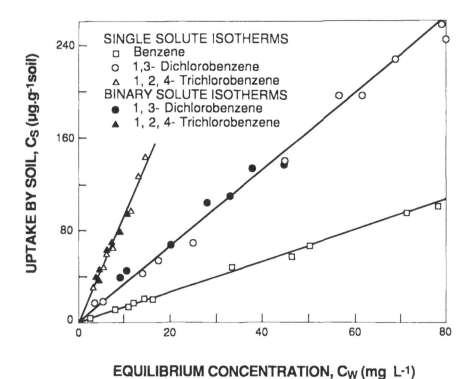

FIGURE 2. Typical soil to water isotherms for benzene, 1,3-dichlorobenzene and 1,2,4-trichloro-benzene as single solutes and binary solutes on a silt loam soil at 20°C. (From Chiou, C. T., Porter, P. E. and Schmedding, D. W., *Environ. Sci. Technol.*, 17, 227, 1983. Copyright American Chemical Society. With permission.)

soils or suspended matter in water runoff. Urban runoff also contains an array of polycyclic aromatic hydrocarbons (PAH), and so on, which are transported from the land to the aquatic areas, e.g., Eganhouse et al.[3], Van Fleet and Quinn.[4] As well as these terrestrial sources, many discharges from industry, sewage treatment plants, and so on, contribute directly to the aquatic load, e.g., Connell.[5] In addition, fugacity-based modeling indicates that with many compounds at equilibrium a substantial proportion finally resides in aquatic sediments.

Aquatic sediments can sorb lipophilic compounds from water, and the equilibrium condition between these two phases is described by an isotherm and a sorption coefficient. The sorption coefficient will be referred to here as K_{sw}, but K_p and K_D are also used in the literature to describe this factor. Thus

$$K_{sw} = C_s/C_w$$

where C_s is the sediment concentration and C_w the water concentration. Examples of the isotherms obtained are illustrated by the results of Chiou et al.[7] as shown Figure 2. At higher concentrations, nonlinearity of these isotherms often occurs. For example, Mingelgrin and Gerstl[6] have reported a number of isotherms for lipophilic compounds which show both positive and negative deviation from linearity. This can be expressed mathematically by introduction of a nonlinearity constant (n) such that

$$C_s = K_{sw}C_w^{1/n}$$

However, in the natural environment the concentrations of lipophilic chemicals are usually low and in most cases n approaches unity. The equation above is commonly described as the Freundlich equation. A linear form can be obtained by taking logs yielding the equation below.

$$\log C_s = \log K_{sw} + 1/n \log C_w$$

Thus, a plot of $\log C_s$ against $\log C_w$ has a slope of $1/n$ and an intercept on the $\log C_s$ axis of $\log K_{sw}$. These relationships have been described by Tinsley,[8] and in addition, this author has described the Langmuir equation which can also be used to describe the sediment-to-water relationship.

The characteristic K_{sw} is a key parameter describing the partition between soil or sediment and water, and varies with different soil or sedimentary types. However, with lipophilic compounds the partition process is substantially controlled by the organic matter or organic carbon content in the sediments, as described by Karickhoff.[9] Other soil characteristics, such as particle size distribution, clay mineral content, pH, cation exchange capacity, and so on, are of much less importance. Thus, partition between sediments and water can be more accurately expressed as a partition between the organic matter in the sediment and water. This means

$$C_{soc} = K_{oc} C_w$$

where C_{soc} is the concentration in sediments expressed in terms of organic carbon, and K_{oc} the corresponding partition coefficient between organic carbon and water. Karickhoff[9] reports that the use of K_{oc} results in values which are much closer to constancy among the different sedimentary types than the use of K_{sw}. Similarly, the sediment-to-water partition coefficient can be expressed in terms of organic matter rather than organic carbon, with similar results being obtained. The expression K_{om} is usually used to describe this characteristic.

The relationships outlined above apply only when equilibrium has been attained. The factors governing the rate of attainment of equilibrium are not well understood, but in many cases lengthy periods may be required. Karickhoff[9] has discussed some of the data which is available on this matter and the possible factors which are involved. This indicates that in some situations the time period over which the sediment-to-water partition process has operated may influence the water concentrations observed, since equilibrium may not have not been reached.

In natural aquatic systems environmental factors may exercise an important influence on the sediment-to-water partition process. For example, evaporation from the surface of a natural water body occurs. Molecules evaporating from the water surface result from movement of molecules at depths near the sediment surface to the water surface where evaporation can occur. Tinsley[10] has reported that the rate of evaporation from different depths of water is governed by the following equation

$$C_t = C_O \exp(-K_L t/d)$$

where C_t is concentration in water at time t, C_O is concentration at time of zero, K_L is the overall liquid exchange constant, t is time period, and d is depth. From this it can be shown that the time period for the evaporation of a constant fraction of compound increases with depth. Thus, with compounds originating from the bottom sediments, a profile of decreasing water concentration with depth may occur. This means that in many natural systems, water concentration values in accord with that expected from the equilibrium equations outlined previously may only occur close to the sediment surface. This situation will be influenced by such factors as turbulence and wind speed at the surface. Thibodeaux[11] has given a detailed analysis of many of the physicochemical factors which influence the exchange of chemicals between water and sediments. Of course, if equilibrium is not attained, the water concentrations generated by compounds sorbed into the bottom sediments will be less than that expected from the relationships previously described.

Biological factors can also be important in some environments. For example, Sondergren and Larsson[12] have found that the generation of gases in anaerobic sediments can aid the distribution of PCBs to the atmosphere. Also, Larsson[13] has found bioturbation to be an important factor in releasing PCBs from sediments to the overlying water.

The sedimentary reservoir of xenobiotic lipophilic compounds can be expected to maintain related concentrations of lipophilic compounds in the overlying water through partition processes. Thus, the water can provide a reasonably continuous supply of lipophilic compounds to aquatic organisms residing in the overlying water mass. This supply could continue over an extended time period since the sediments can contain relatively high amounts of lipophilic compounds (see Table 1). Also, it is noteworthy that proportions of these compounds can be released from the water to the atmosphere and so contribute through this route to terrestrial biota (see Figure 1).

Within aquatic sediments, similar partition processes to those described above for the bottom sediments and the water mass would be expected. The interstitial water concentration of lipophilic compounds would be reflected by the Freundlich equation modified for organic carbon and the relatively low concentrations expected in this environment. This would mean that $C_{soc} = K_{oc} C_w$. In this environment, however, the processes of turbulence, depth of water, and wind speed at the surface, would be expected to be of greatly reduced importance. The times to establish equilibrium may be comparatively short. Thus, a ready supply of lipophilic compounds dependent on replenishment from the reservoir sorbed onto the sediments would be available in the interstitial water. Shaw and Connell[14] have suggested that this material could be bioconcentrated by aquatic benthic infauna. In more recent work, Connell et al.[41] have shown that bioconcentration of lipophilic compounds from sediment by infauna is similar to bioconcentration by fish.

The previous discussion indicates that the sediments are a major repository for xenobiotic lipophilic compounds and the overlying water, in equilibrium, is the principal route of uptake of these substances by aquatic biota. However, some organisms, particularly higher trophic level aquatic biota, obtain at least a proportion of their xenobiotic lipophilic chemical load from food sources (see Figure 1). But the water to low trophic level aquatic biota system controls the level of these chemicals in food, and the direct partitioning between water and the higher trophic biota, irrespective of the source of the chemical, also exercises some control. Thus, the actual levels which occur in higher trophic level aquatic biota would be expected to be influenced by environmental partition processes.

C. THE SOIL-TO-ATMOSPHERE SYSTEM

A wide variety of lipophilic chemicals are used and distributed to the terrestrial atmosphere or terrestrial soil. For example, many pesticides are used in agriculture by spraying onto crops, as well as by application in the soil. Wheatley[15] has concluded that evaporation from soil, which can be assisted by soil erosion, is the most important source of atmospheric pesticides. Miller and Connell[16] have described how petroleum hydrocarbons and PAHs are discharged into the atmosphere in comparatively large quantities through the burning of fuels. Also, Nisbet and Sarofilm[17] have pointed out how the incineration of plastics results in the release of polychloro biphenyls (PCBs) to the atmosphere. Some examples of compounds which have been measured in the atmosphere of different regions are shown in Table 2, and indicate that a range of concentrations and wide range of types of compounds can be expected.

Compounds can be discharged to the atmosphere as vapors or as aerosols and particulates. Seiber et al.[18] have reported that much of this material is sorbed or deposited in soil close

TABLE 2
Examples of the Occurrence of Lipophilic Compounds in the Atmosphere

Compound	Location	Concentration range (μg m^{-3})
DDT	Forests (treated)	540—171,000
Endrin	Vegetable fields (treated)	20—90
BHC	Forests (treated)	2600—4600
DDT	Community (usage area)	0.0001—0.022
Chlordane	Usage areas	0.0001—0.006
Aldrin	Florida	0.0001—0.004
DDT and metabolites	London	0.0038—0.0071
DDT	U.S.	0.0001—1.56
DDT and metabolites	Barbados	4.3×10^{-9}—4.5×10^{-8}
PCBs	U.S., Canada, Sweden, Japan	10—50,000

Compiled from Connell, D. W. and Miller, G. J., *Chemistry and Ecotoxicology of Pollution*, John Wiley & Sons, New York, 1984, 169.

to its origin and the concentrations are higher in the atmosphere at this location also. The data in Table 2 indicate this general pattern, but also indicate that air in many urban situations contains low concentrations of agricultural pesticides.

The rate of entry to the atmosphere and the distance of movement are principally dependent on the vapor pressure and meteorological conditions. In some cases, movements on a global scale have been suggested. For example, pesticides measured in the atmosphere in Barbados (Table 2) have been identified as originating in Europe or Africa by Riseborough et al.[19] In fact, Wheatly[15] has suggested that the atmosphere is a major transport route for the distribution of pesticides and possibly other lipophilic chemicals on a global scale.

Terrestrial soil is a major final respository for many lipophilic chemicals after discharge into the terrestrial environment (see Table 1). Generally, the sorption of chemicals on soil is influenced by a similar range of properties to those that influence sorption onto sediments. Factors that have been found to have a limited influence are particle size distribution, clay mineral content, pH, cation exchange capacity, and so on. But similarly to aquatic sediments with lipophilic compounds, the organic matter present, expressed as organic carbon, is a key parameter. This organic matter consists principally of remains of plant and animal debris, as well as micro-organisms which may utilize these components. The soil-to-air partition process is characterized by a partition coefficient, expressed as K_{SOA}, thus

$$K_{SOA} = C_{SO}/C_A$$

where C_A is the concentration in air, and C_{SO} the concentration in soil.

However, the soil-to-atmosphere process is complex and is influenced by such factors as presence of water, depth of soil involved, and so on. Nevertheless, the air-to-soil partition coefficient is related to the Henry's Law Constant and the sediment-to-water partition coefficient. Of course, it can be determined experimentally by directly measuring C_{SO} and C_A under controlled conditions. This parameter characterizes the distribution of a chemical between the atmosphere and soil, and can be used in fugacity-based calculations of the distributions and concentrations of a chemical likely to reside in soil and the atmosphere at equilibrium.

Thibodeaux[20] has reviewed the methods used to calculate the exchange rates in the air-to-soil system. These rates are strongly influenced by the micrometeorological conditions in the air layer immediately above the soil. Thermal turbulance above the air-to-soil surface has a strong influence on these processes. Since equilibrium with the atmosphere is unlikely to be established due to the dynamic nature of the atmosphere, these exchange rates are of considerable importance in evaluating the persistence of a chemical in soil.

Soil organisms, for example, earthworms, and biota consuming some soil in their diet, would be expected to accumulate xenobiotic chemicals directly from this source. In addition, the soil can provide a reservoir to replenish atmospheric concentrations of compounds which may be bioaccumulated by terrestrial biota from this source.

D. THE ATMOSPHERE-TO-TERRESTRIAL BIOTA PROCESSES

Connell and Hawker[21] have utilized a theoretically derived partition coefficient for the vegetation-to-air partition coefficient at equilibrium, which has enabled them to include a vegetation phase in their fugacity model. There is a lack of data on this process available in the literature. Utilizing an urban Australian environment, the distributions shown in Table 3 were obtained. These data indicate that a major proportion of these typical xenobiotic lipophilic compounds is often in the atmosphere-to-soil system. In some cases, however, the atmosphere contains a very large proportion of the compound at equilibrium. But due to the large size of this phase and its dynamic nature, not expressed in this model, the concentrations can be relatively low. The concentrations in vegetation obtained by Connell and Hawker[21] were calculated for foliage only and are among the highest concentrations listed in Table 3. In some environments the vegetation phase could be expected to be very large, and thus take up a significant proportion of the total chemical available. These results indicate that the soil-to-atmosphere transfer route leading to terrestrial vegetation may be significant in many environments.

The consumption of terrestrial vegetation by low trophic level terrestrial biota could be expected to result in the accumulation of residues in these biota. In addition, at least a proportion of these residues would be passed onto higher trophic level terrestrial biota through food consumption patterns in the food web (see Figure 1).

With higher trophic level terrestrial biota and low trophic level terrestrial biota, there is the possibility of accumulation of chemicals by partitioning between biota lipids and the atmosphere (see Figure 1). In many cases, the atmospheric concentrations will be exceptionally low due to the dynamic nature of this phase and the low concentrations of many chemicals which finally reside in the phase. In these cases the concentrations may be so low that significant bioaccumulation may not occur. Thus, the importance of this pathway will depend on the particular chemical involved and the environmental conditions.

III. DEFINITION OF BIOCONCENTRATION AND BIOMAGNIFICATION

The transfer of xenobiotic chemicals from various sources within the environment to organisms, resulting in a generally higher concentration within the organism than the source, is described by various terms. Many of the terms used attempt to give an indication of the actual mechanism involved in the transfer process. For example, the occurrence of apparent increases in the concentrations of a chemical with trophic level has been described as food chain transmission (Hunt and Bischoff[22]), transfer along a food chain (Bryan[23]), ecological magnification (Metcalf et al.[24]), food chain magnification (Tinsley[25]), and a variety of other terms reported by Shaw and Connell[26] (e.g., bioamplification, trophic contamination, trophic magnification). A diagrammatic representation of the movement of xenobiotic chemicals from sources within the environment to biota in the natural environment is shown in Figure 1. This indicates that two basic processes are involved, first direct transfer from the abiotic environment to an organism, and second, transfer within food to a consumer. In both of these cases subsequent increases in concentration above that in the source are considered to result in bioaccumulation.

The process of direct transfer from water to an organism is described in this publication as "bioconcentration". This terminology is now widely accepted as appropriate for this

TABLE 3
Calculated Potential Distribution and Concentration of Some Representative Chemicals in a Typical Australian Urban Environment

	Pentachlorophenol		Phenanthrene		Tetrachloroethene		DDT		Hexachlorobiphenyl	
	μg/kg[a]	% Moles[b]	μg/kg	% Moles	μg/kg	% Moles	μg/kg	% Moles	μg/kg	% Moles
Air	6×10^{-5}	6×10^{-3}	2×10^{-3}	0.32	0.5	82.7	5×10^{-6}	3×10^{-4}	1×10^{-4}	8×10^{-3}
Water	6×10^{-5}	7×10^{-4}	1×10^{-4}	2×10^{-3}	2×10^{-3}	0.03	5×10^{-6}	4×10^{-5}	2×10^{-7}	2×10^{-6}
Soil	35.5	86.8	23.6	86.4	3.8	14.9	47	86.5	47.9	86.4
Suspended sediments	107	0.18	71	0.18	11.4	0.03	141	0.18	144	0.18
Sediment	107	12.7	71	12.6	11.4	2.17	141	12.6	144	12.6
Aquatic biota	282	2×10^{-3}	187	2×10^{-3}	30	3×10^{-4}	376	2×10^{-3}	379	2×10^{-3}
Vegetation	113	0.62	75	0.61	11.9	0.10	150	0.61	152	0.61

[a] Concentration in a single phase.

[b] Percentage distribution of the compound between all phases.

From Connell, D. W. and Hawker, D. W., *Chem. Aust.*, 53, 427, 1986. Copyright Royal Australian Chemical Institute. With permission.

process. Thus, in Figure 1 the transfer from water to low trophic level aquatic biota, higher trophic aquatic biota, and to filter feeders within the aquatic environment is described as bioconcentration. In addition, bioconcentration also occurs with the transfer of chemicals from the interstitial water to infauna. In a broader sense, the transfer of xenobiotic chemicals from air to terrestrial vegetation, low trophic level terrestrial biota, and higher trophic level biota is an analagous process, but the inclusion of these processes under the term bioconcentration is inappropriate. However, perhaps "terrestrial bioconcentration" is a suitable terminology to use to distinguish this process from bioconcentration in the aquatic environment.

The bioconcentration process can be seen as a partition process between the ambient water environment and aquatic biota. This is illustrated in Figure 1 by double headed arrows for those processes which can be seen as having a direct partition mechanism, and characterized by K_B where K_B is equal to C_B/C_W for a compound at equilibrium (C_B is the concentration in biota and C_W the concentration in water). Similarly, terrestrial bioconcentration for a particular compound can be characterized by an analogous bioconcentration factor equal to C_B/C_A, where C_A is equal to the concentration in the atmosphere at equilibrium.

The tranfer of a xenobiotic chemical from food to an organism is described here as biomagnification. In Figure 1 chemical transfer from low trophic level aquatic biota to higher trophic level aquatic biota in the aquatic environment, and from terrestrial vegetation to low trophic level terrestrial biota, and low trophic level terrestrial biota to higher level trophic terrestrial biota in the terrestrial environment, can be seen as biomagnification. In addition, transfer between the aquatic and terrestrial environments, for example from infauna, low and higher trophic aquatic biota to higher trophic level terrestrial biota, is a biomagnification process. Biomagnification, as it is used here, does not necessarily imply a sequential increase in concentration with trophic level, but that a transfer from food to consumer of a xenobiotic chemical has occurred over one or more trophic levels. This process is not characterized by a simple equilibrium as has proved adequate for bioconcentration, and this is indicated by single headed arrows in Figure 1. Thus, a partition coefficient, indicating an equilibrium situation, is not appropriate. However, the evidence suggests that more complex partition processes are probably involved in biomagnification.

In many cases, the mechanism of transfer is unknown or does not need to be specified. In these cases the term bioaccumulation is used. Thus, this term simply means that a tranfer has occurred and the concentration in the organism is higher than the concentration in the source.

IV. OPERATION OF BIOCONCENTRATION AND BIOMAGNIFICATION IN NATURAL SYSTEMS

A. THE BIOMAGNIFICATION PROCESS

Interest in the bioaccumulation of environmental xenobiotic compounds started in the 1950s and 1960s. During this period widespread occurrence of chlorinated hydrocarbons used in agriculture, particularly DDT, were detected in the natural environment (Rudd and Genelly,[27] Zitko et al.,[28] Robinson et al.[29]). In some cases, a relationship between biotic concentration and trophic level occurred, as outlined by Woodwell et al.[30] and Woodwell,[31] which led these early workers to propose a related mechanism. The mechanism proposed was due to the differences in metabolic behavior of the insecticides as compared to the food materials. Respiration results in a degradation of the food substances by an organism, but the insecticides are resistant to this process; they remain and are therefore concentrated. This concentration process occurs at each step in a food chain, resulting in the observed relationship between trophic level and biotic insecticide concentration.

More recent investigations have demonstrated that while the stability of the xenobiotic compounds is a key factor in their bioaccumulation, there are a number of additional factors involved which also exert major effects. For example, complete uptake of all of the available xenobiotic compound in food rarely occurs. Connell[32] has collated data on the efficiency of uptake of xenobiotic compounds in food by organisms, which indicate a high degree of variability. This variability may result from the different species involved in a food chain or with different types of compound. Also, Hamelink et al.[33] and Moriarity[34] have demonstrated that in some cases, residues in the food chain can be explained by bioconcentration alone.

The available evidence indicates that in aquatic ecosystems higher trophic level aquatic biota may acquire xenobiotic compounds by biomagnification and bioconcentration, as indicated in Figure 1. In terrestrial ecosystems, the air to lower trophic level terrestrial biota and higher trophic level terrestrial biota partition processes are usually not of significance due to the extremely low concentrations of compounds which occur in the atmosphere. Thus, biomagnification in these systems is the predominant route of bioaccumulation.

B. THE BIOCONCENTRATION PROCESS

Bioconcentration in aquatic organisms involves transfer of the xenobiotic compound from water to the gills, or body surface, then to the circulatory fluid which is followed by either metabolism or storage. The uptake path is similar to the path of uptake of dissolved oxygen from water. Connell[32] has reported that the surfaces involved, the respiratory surfaces, must be permeable to oxygen, and thus other compounds may also utilize the same uptake route. Also, the organism must have a mechanism for the internal distribution of oxygen, which will therefore allow the distribution of absorbed xenobiotic compounds as well.

Terrestrial bioconcentration principally involves plants, and quite different mechanisms are involved from those in aquatic bioconcentration. Riederer and Schonherr[35] have shown that plant cuticle has a high capacity to bioconcentrate lipophilic compounds. Xenobiotic lipophilic chemicals can reach the plant foliar surface on particles or as vapor and be absorbed into cuticle lipids and subsequently pass into plant cells, as described by Norris.[36] The volatilization of compounds from the adjacent soil is often the major source of chemicals bioconcentrated by plants through foliar uptake (Topp et al.[37]).

Another route for the bioconcentration of xenobiotic chemicals by terrestrial vegetation is through the root system from soil. Topp et al.[37] report that while the foliar pathway has been demonstrated to be more important in some cases, under field conditions the root pathway is more important.

C. SIGNIFICANCE OF BIOCONCENTRATION AND BIOMAGNIFICATION WITH AQUATIC ORGANISMS

Both bioconcentration and biomagnification must operate with most aquatic organisms, as indicated in Figure 1. But these mechanisms involve different pathways to the organism and may have different control characteristics. Thus, a knowledge of the significance of these pathways in aquatic organisms may be helpful in understanding the bioaccumulation process.

Air breathing aquatic animals, such as seals, whales, dolphins, and so on, as well as semi-aquatic species, such as aquatic birds, lack an organism to water exchange interface. thus, the bioconcentration mechanism cannot operate, and biomagnification is the sole process by which these animals acquire xenobiotic compounds. In this sense these organisms are terrestrial organisms, and more appropriately included with higher trophic level terrestrial biota in Figure 1.

On the other hand, autotrophic organisms, such as phytoplankton and some bacteria, draw their food and oxygen requirements from dissolved components in the water mass. With these organisms bioconcentration must operate, since the only xenobiotic compounds

TABLE 4
DDT Residues Data Reported in 1967[a]

Area	Biota	Tissue	Concentration (ppm)
California	Plankton	—	5.3
California	Bass	Edible flesh	4.138
California	Grebes	Visceral fat	Up to 1,600
Montana	Robin	Whole body	6.8—13.9
Wisconsin	Crustacea	—	.41
Wisconsin	Chub	Whole body	4.52
Wisconsin	Gull	Brain	20.8
Missouri	Bald Eagle	Eggs	1.1—5.6
Connecticut	Osprey	Eggs	6.5
Florida	Dolphin	Blubber	About 220
Canada	Woodcock	Whole body	1.7
Antarctica	Penguin	Fat	.015—.18
Antarctica	Seal	Fat	.042—.12
Scotland	Eagle	Eggs	1.18
New Zealand	Trout	Whole body	.6—.8

[a] Including the derivatives DDD and DDE as well as DDT itself (data selected from hundreds of reports)

From Woodwell, G. M., *Scientific American*, 216, 24, 1967. Copyright *Scientific American*. With permission.

available are those dissolved in the water. These substances bioconcentrate within the organism by the oxygen uptake route i.e., through the outer membrane surface. In fact, Baughman and Paris[38] have reported that the bioaccumulation of lipophilic compounds by a range of microbes could be explained by the organism-to-water partition process.

Many investigations[27-30] of DDT and related compounds in the natural environment were conducted from the 1950s onwards. Generally, the carnivorous birds contained the highest concentrations, and an apparent increase in concentration with trophic level was observed with some sets of data. This was cited as evidence of food chain biomagnification, with repeated increases in biotic concentration occurring at each step in the food chain. For example, in 1969 Woodwell[31] collated much of the information then available on DDT residues, as shown in Table 4. But in many of the following investigations, illustrated by Shaw and Connell[39] and Robinson,[29] there was no clear evidence of an increase in the concentration of xenobiotic chemicals with trophic level. In 1971 Hamelink et al.[33] suggested that the concentration of xenobiotic lipophilic compounds in aquatic ecosystems could be explained by organism-to-water partitioning. Also, Griesbach[40] was able to describe the residues in aquatic ecosystems on the basis of an allometric mathematical model in which trophic level was not a factor.

At present, a great deal of further information is available on such factors as organism uptake efficiencies from food, influence of gross rate of bioaccumulation, rates of uptake and loss of lipophilic compounds, and so on. However, a clear interpretation of the relative significance of bioconcentration and biomagnification in aquatic systems has not emerged. Connell[32] has concluded that:

1.   Biomagnification operates with air breathing aquatic organisms.
2.   Bioconcentration operates with autotrophic organisms, such as phytoplankton and some bacteria.
3.   Both mechanisms vary in importance in various other biotic groups depending on a variety of factors, but in general, bioconcentration is of greater significance.

4. Biomagnification is mostly likely to occur with persistent compounds having log K_{ow} greater than 5, and with organisms having long lives and probably among the top predators.

REFERENCES

1. **Mackay, D. and Patterson, S.**, Calculating fugacity, *Environ. Sci. Technol.*, 15, 106, 1981.
2. **Mackay, D. and Patterson, S.**, Fugacity revisited, *Environ. Sci. Technol.*, 16, 654A, 1982.
3. **Eganhouse, R. F., Simoneit, B. R. T., and Kaplan, J. R.**, Extractable organic matter in urban storm water runoff. 2. Molecular Characterisation, *Environ. Sci. Technol.*, 15, 315, 1981.
4. **Van Fleet, E. S. and Quinn, J. G.**, Input and fate of petroleum hydrocarbons entering the Providence River and Upper Narraganett Bay from wastewater effluents, *Enviorn. Sci. Technol.*, 11, 1087, 1977.
5. **Connell, D. W.**, An approximate petroleum hydrocarbon budget for the Hudson Raritan Estuary, New York, *Mar. Pollut. Bull.*, 13, 89, 1982.
6. **Mingelgrin, U. and Gerstl, Z.**, Re-evaluation of partitioning as mechanism of non-ionic chemical adsorption in soils, *J. Environ. Qual.*, 12, 1, 1983.
7. **Chiou, C. T., Porter, P. E., and Schmedding, D. W.**, Partition equilibria of non-ionic organic compounds between soil organic matter and water, *Environ. Sci. Technol.*, 17, 227, 1983.
8. **Tinsley, I. J.**, *Chemical Concepts in Pollutant Behaviour*, John Wiley & Son, New York, 1979, 12.
9. **Karickhoff, S. W.**, Sorption phenomena, in *Environmental Exposure from Chemicals*, Vol. I., Neely, W. B. and Blau, G. E., Eds. CRC Press, Boca Raton, 1985, 49.
10. **Tinsley, I. J.**, *Chemical Concepts in Pollutant Behaviour*, John Wiley & Son, New York, 1979, 57.
11. **Thibodeaux, I. J.**, *Chemodynamics—Environmental Movement of Chemicals in Air, Water and Soil*, John Wiley & Sons, New York, 1979, 225.
12. **Sondergren, A. and Larsson, P.**, Transport of PCBs in aquatic laboratory model ecosystems from sediment to the atmosphere via the surface microlayer, *Ambio*, 11, 43, 1982.
13. **Larsson, P.**, Transport of ^{14}C-labelled PCB compounds from sediment to water and from water to air in laboratory model systems, *Water Res.*, 17, 1317, 1983.
14. **Shaw, G. R. and Connell, D. W.**, Comparative kinetics for bioaccumulation of polychlorinated biphenyls by the polychaete *(Capitella Capitata)* and fish *(Mugil Cephalus, Ecotoxicol. Environ. Saf.*, 13, 84, 1987.
15. **Wheatley, G. A.**, Pesticides in the atmosphere, in *Environmental Pollution by Pesticides*, Edwards, C. A., Ed., Plenum Press, London, 1973, 365.
16. **Miller, G. J. and Connell, D. W.**, Global production and fluxes of petroleum and recent hydrocarbons, *Int. J. Environ. Stud.*, 19, 273, 1982.
17. **Nisbet, C. T. and Sarofilm, A. F.**, Rates and routes PCBs in the environment, *Environ. Health Perspect.*, 1, 21, 1972.
18. **Seiber, J. N., Woodrow, J. E., Shafik, T. A., and Enos, H. F.**, Determination of pesticides and their transformation products in air, in *Environmental Dynamics of Pesiticides*, Harque, R. and Freed, V. H., Eds., Plenum Press, New York, 1975, 17.
19. **Risebrough, R. W., Huggett, J. J. R., Griffin, J. J., and Goldberg, E. D.**, Pesticides: Trans-atlantic movements of in the North East Trades, *Science*, 159, 1233, 1968.
20. **Thibodeaux, L. J.**, *Chemodynamics—Environmental Movement of Chemicals in Air, Water and Soil*, Wiley & Sons, New York, 1979, 139.
21. **Connell, D. W. and Hawker, D. W.**, Predicting the distribution of persistent organic chemicals in the environment, *Chem. Aust.*, 53, 428, 1986.
22. **Hunt, E. G. and Bischoff, A. I.**, Inimical effects on wildlife of periodic DDD applications to clear lake, *Calif. Fish Game*, 46, 91, 1960.
23. **Bryan, G. W.**, Bioaccumulation of marine pollutants, *Philos. Trans. R. Soc. London*, B286, 483, 1979.
24. **Metcalf, R. L., Sangha, G. K., and Kapoor, I. P.**, Model ecosystem for the evaluation of pesticide biodegradability and ecological magnification, *Environ. Sci. Technol.*, 5, 709, 1971.
25. **Tinsley, J.**, *Chemical Concepts in Pollutant Behaviour*, John Wiley & Sons, New York, 1979, 191.
26. **Shaw, G. R. and Connell, D. W.**, Factors controlling PCB's in food chains, in *PCB's and the Environment*, Waid, J. S., Ed., CRC Press, Boca Raton, FL, 135, 1986.
27. **Rudd, R. L. and Genelly, R. E.**, Pesticides: their use and toxicity in relation to wildlife, *Calif. Fish Game*, 7, 1956.
28. **Zitko, V., Hutzinger, O., and Choui, P. M. K.**, Contamination of the Bay of Fundy — Gulf of Marine area with polychlorinated biphenyls, polychlorinated terphenyls, chlorinated bibenzodioxins and dibenzofurans, *Environ. Health Perspect.*, 1, 47, 1972.

29. **Robinson, J., Richardson, A., Crabtree, A. N., Coulson, J. C., and Potts, G. R.,** Organochlorine residues in marine organisms, *Nature,* 214, 1307, 1967.
30. **Woodwell, G. M., Wurster, C. F., and Isaacson, P. A.,** DDT residues in an east coast estuary. A case of biological concentration of a persistent insecticide, *Science,* 156, 821, 1967.
31. **Woodwell, G. M.,** Toxic substances and ecological cycles, *Sci. Am.,* 216, 24, 1967.
32. **Connell, D. W.,** Bioaccumulation behaviour of persistent organic chemicals by aquatic organisms, *Rev. Environ. Contam. Toxicol.,* 101, 117, 1988.
33. **Hamelink, J. L., Waybrant, R. C., and Ball, C.,** A proposal: exchange equilibria control the degree chlorinated hydrocarbons are biologically magnified in lentic environments, *Trans. Am. Fish. Soc.,* 100, 207, 1971.
34. **Moriarty, F.,** Exposure and residues, in *Organochlorine Insecticides: Persistent Organic Pollutants,* Moriarty, F. Ed., Academic Press, London, 1975, 29.
35. **Riederer, M. and Schonherr, J.,** Accumulation and transport of 2,4-dichlorophenoxyacetic acid in plant cuticles, *Ecotoxicol. Environ. Saf.,* 9, 196, 1985.
36. **Norris, R. F.,** Penetration of 2,4-D in relation to cuticle thickness, *Am. J. Bot.,* 61, 74, 1974.
37. **Topp, E., Scheunert, I., Attar, A., and Korte, F.,** Factors affecting the uptake of ^{14}C-labelled organic chemicals by plants from soil, *Ecotoxicol. Environ. Saf.,* 11, 219, 1986.
38. **Baughman, G. L. and Paris, D. F.,** Microbial bioconcentration of organic pollutants from aquatic systems—a critical review, *CRC Crit. Rev. Microbiol.,* Jan., 205, 1981.
39. **Shaw, G. R. and Connell, D. W.,** Factors influencing concentration of polychlorinated biphenyls in organisms from an estuarine system, *Aust. J. Mar. Freshwater Res.,* 33, 1057, 1982.
40. **Griesbach, S., Peters, R. H., and Youakim, S.,** An allometric model for pesticide bioaccumulation, *Can. J. Fish. Aquat. Sci.,* 39, 727, 1982.
41. **Connell, D. W., Bowman, M., and Hawker, D. W.,** Bioconcentration of chlorinated hydrocarbons from sediment by oligochaetes, *Ecotoxicol. Environ. Saf.,* 16, 293, 1988.

Chapter 5

THE PARTITION MECHANISM

Darryl W. Hawker

TABLE OF CONTENTS

I. INTRODUCTION

For many persistent, xenobiotic, hydrophobic compounds bioconcentration can be satisfactorily treated as a physical-chemical partitioning process between the lipid phases of an aquatic organism and the surrounding aqueous medium.[1,2] For example, bioconcentration factors (K_B) for fish have generally been shown to be related to the n-octanol/water partition coefficient K_{OW}.[3-6] Abiotic partition coefficients, particularly K_{OW}, are useful predictors of environmental partitioning behavior in general. The measurement of K_{OW} for new chemicals is, as noted by Watarai,[7] officially recommended in the OECD Chemical Testing Program. Both soil/water and sediment/water partition coefficients and aqueous solubility have been correlated with K_{OW}.[8,9]

The first systematic studies on the distribution of a solute between two phases were carried out towards the end of the 19th century. These early investigations revealed that the ratio of the concentrations of the solute in two immiscible solvents was roughly constant and independent of the solvent volumes employed. Later, Nernst found that this ratio was only constant if the solute underwent no change, such as dissociation or association, on partitioning.[10] He also treated partitioning thermodynamically by considering it as an equilibrium process. The partial molar free energy for a solute in solvent A ($\overline{G}_A$) is given by

$$\overline{G}_A = \overline{G}_A^\circ + RT \ln a$$

where $\overline{G}_A^\circ$ is a standard reference free energy for the solute, R the universal gas constant, a the activity, and T the temperature (Kelvin). At equilibrium the partial molar free energies in each phase (A and B) are equal, hence

$$\ln \frac{a_A}{a_B} = \frac{\overline{G}_B^\circ - \overline{G}_A^\circ}{RT} \tag{1}$$

For a given temperature, $\overline{G}_A^\circ$, $\overline{G}_B^\circ$ and R are constants, and therefore

$$\frac{a_A}{a_B} = K = \text{constant} \tag{2}$$

Equation 2 is a mathematical statement of Nernst's distribution law. When the solutions are dilute, or when the solute is an ideal one, the activity is approximately equal to the concentration in each phase, and Equation 2 becomes

$$\frac{C_A}{C_B} = K$$

The dimensionless constant, K, is referred to as the partition coefficient of the solute between the two solvents A and B.[11] From Equation 1, it is apparent that the partition coefficient has a temperature dependence. Chiou et al.[12] have suggested that the temperature effect on the partition coefficient is generally about 0.01 log units per degree around room temperature.

An expression describing the partition coefficient may also be derived from the viewpoint of chemical potential. The chemical potential (μ) of a solute distributed between two phases A and B, is given by

$$\mu_A = \mu^0 + RT \ln \gamma_A \phi_A$$

$$\mu_B = \mu^0 + RT \, \ell n \, \gamma_B \, \phi_B$$

where μ^0 is the standard chemical potential of the solute and γ and ϕ the phase-specific solute volume fraction activity coefficient and volume fraction, respectively. Interphase solute transfer occurs spontaneously from the phase with the higher chemical potential, until μ is constant throughout the system,[13] when

$$\gamma_A \, \phi_A = \gamma_B \, \phi_B$$

Since $\phi_i = C_i V_s^i$ where C_i is the solute concentration in a solvent $_{ii}$ and V_s^i is the solute molar volume in solvent i (a constant),[14] then

$$K_{AB} = \frac{C_A}{C_B} = \frac{\gamma_B V_s^B}{\gamma_A V_s^A} \tag{3}$$

Another alternative expression for K can be obtained by using Raoult's law convention. At equilibrium

$$\gamma_A^* \, X_A = \gamma_B^* \, X_B$$

where γ_A^* and γ_B^* are solute mole fraction activity coefficients for the respective phases, and X_A and X_B solute mole fractions.[15] Phase concentration is roughly equal to X_A/V_A, where V_A is the molar volume of solvent A, and therefore

$$K_{AB} = \frac{C_A}{C_B} = \frac{\gamma_B^* \, V_B}{\gamma_A^* \, V_A} \tag{4}$$

Chemical potential is expressed in units of energy per mole, and is often awkward to use and conceptually difficult to grasp. As a result, G. N. Lewis introduced fugacity as a better measure of equilibrium between phases.[16,17] Fugacity has units of pressure and is related to chemical potential by

$$\mu = \mu^0 + RT \, \ell n \, f$$

where f is the fugacity of a solute in a given phase. Fugacity can be thought of as the escaping tendency of a solute, and transfer of a solute occurs spontaneously from the phase with the highest free energy, chemical potential, or fugacity, until these functions are equal in each phase, when the system is at equilibrium.

Treatments such as those presented above are based on the assumption of immiscibility of solvents, and use pure phase physical constants. In practice, most solvents are mutually soluble in each other to a certain degree, and constants taking into account this solution should be strictly employed. Often though, use of pure phase constants results in a reasonable approximation.

II. NATURE OF THE PARTITIONING MECHANISM

A. PRINCIPLES OF PARTITIONING

The partitioning of a non-ionizable solute between two essentially immiscible solvents can be initially considered analogous to a two step process in which the solute is first transferred from solution to the gaseous phase or state, and then to solution again in the other solvent.[18] This process is illustrated diagrammatically in Figure 1 (Scheme 1). The

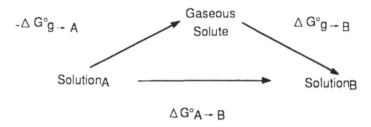

SCHEME 1: PARTITIONING

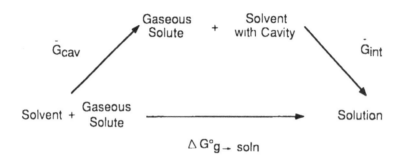

SCHEME 2 : GAS-TO-SOLUTION PROCESS

FIGURE 1.   A stepwise analysis of the partitioning process. (From Dunn, W. J., Block, J. H. and Pearlman, R. S., *Partition Coefficient — Determination and Estimation*, Chap. 1., 1986. Copyright Pergamon Press, Elmsford, NY. With permission.)

free energy change for this overall process is of course related to the partition coefficient by

$$RT \ln K_{AB} = \Delta G^\circ_{g \to A} - \Delta G^\circ_{g \to B}$$

where K_{AB} is the partition coefficient between A and B, and $\Delta G^\circ_{g \to soln}$ is the free energy change associated with the process of transfer of solute from solution to gaseous phase.

Extending this concept further, each step in Scheme I can be regarded as consisting of two steps. These are the formation of a cavity of sufficient size to accommodate the solute, and the transfer of a solute molecule from the gaseous phase into the cavity. This situation is illustrated in Scheme 2 of Figure 1. Thermodynamically, the free energy change between gaseous phase and solution can be divided into

$$\Delta G^\circ_{g \to soln} = \overline{G}_{cav} + \overline{G}_{int}$$

where $\overline{G}_{cav}$ is the partial free energy for the cavity formation process, and $\overline{G}_{int}$ represents the partial free energy for interaction between solutes and solvents.[18] Thus, the partition coefficient is related to the difference in cavity formation free energies, and the difference in solute-solvent interaction free energies for the two solvents. Watarai et al.[7] reached a similar conclusion by considering that the free energy for the partition process was represented by the difference in solution free energies of solute in both phases, and that the free energy of solution could be expressed by

TABLE 1
Contribution of Energy Terms (kJ Mol^{-1}) to the Partitioning of Benzene Between Heptane and Water, and Between n-Octanol and Water

Solvent	$\overline{G}_{cav}$	$\overline{G}_{dis}$	$\overline{G}_{ind}$
Water	37.8	−39.4	−5.06
Heptane	18.2	−38.6	0
n-Octanol	27.2	−42.4	−1.17

From Watarai, H., Tanaka, M., and Suzuki, N., *Anal. Chem.*, 54, 702, 1982. Copyright American Chemical Society. With permission.

$$\overline{G}_{cav} + \overline{G}_{int} + RT \, \ell n \, (RT/10^3)$$

From scaled particle theory, the cavity formation energy may be described by

$$G_{cav} = RT \, \ell n \, (1 - y) + RT \frac{3y}{1 - y} \left(\frac{\sigma_2}{\sigma_1}\right)$$
$$+ RT \left[\frac{3y}{1 - y} + \frac{9}{2} \left(\frac{y}{1 - y}\right)^2\right]\left(\frac{\sigma_2}{\sigma_1}\right)^2 + \frac{NyP}{\rho}\left(\frac{\sigma_2}{\sigma_1}\right)^3$$

where ρ is the density of the solvent, σ is the molecular diameter, subscripts 1 and 2 refer to solvent and solute, respectively, P is the pressure, and y is the compactness factor ($= \pi(\sigma_1)^3/6$).

Similarly, the interaction term can be divided into terms representing dipole-dipole, inductive and dispersion (London) energies,

$$\overline{G}_{int} = \overline{G}_{dip} + \overline{G}_{ind} + \overline{G}_{dis} = -N(16/3) \, 4(\epsilon_{12} \, \sigma_{12}^6) \frac{\pi\rho}{6\sigma_{12}^3}$$
$$- 8N(m_1^2\alpha_2 + m_2^2\alpha_1 + 2/3 \, m_1^2 \, m_2^2/kT) \frac{\pi\rho}{6\sigma_{12}^3}$$

where $\sigma_{12} = (\sigma_1 + \sigma_2)/2$, $\epsilon_{12} = (\epsilon_1\epsilon_2)^{0.5}$ and ϵ, m, and α correspond to the Lennard-Jones energy parameter, the dipole moment, and the molecular polarizability, respectively.[19]

To determine which factors govern partition coefficients of nonpolar solutes, all energy terms have been quantitatively evaluated for the partitioning of benzene between water and octanol, and between water and heptane. The contribution of the dipole-dipole interaction ($\overline{G}_{dip}$) is essentially zero because the dipole moment of benzene is zero. The results of these calculations are presented in Table 1. The cavity formation energy quantifies the free energy input necessary to overcome the solvent-solvent cohesive interactions, and is large and positive for all solvents. It requires most energy to form a cavity in water, and least to form a cavity in heptane, which is the solvent most compatible with the solute, benzene. The dispersion (London) energy term is negative and has a small solvent influence, while the induction term seems to have little effect on the partitioning behavior of benzene. On the basis of these results, it can be concluded that the magnitude of the partition coefficient is principally determined by the difference in cavity formation energies in this instance.

Recently, partition coefficients have been estimated with good precision on the basis of relationships, including a term measuring cavity formation energy requirements and solvatochromatic parameters describing solute dipolarity/polarizability, hydrogen bond donor acidity, and hydrogen bond acceptor basicity.[20] Quantitative estimates of the relative importance of cavity formation and solute-solvent interaction contributions to partitioning have also been made by other methods. Intuitively, the solvent cavity dimensions are related to the size of the solute. For a macroscopic spherical cavity*

$$\overline{G}_{cav} = 4\pi r^2 \cdot \Gamma$$

where Γ is the interfacial tension. On a molecular level, solutes are not always spherical. It has been suggested[18,21] that for a nonspherical cavity on this scale

$$\overline{G}_{cav} = k_{cav} \cdot TSA \tag{5}$$

where k_{cav} is an empirical proportionality constant, and TSA is the total molecular surface area (or some related measure) of the solute. This approach has been criticized by Hermann,[22] who argues that $\overline{G}_{cav}$ should be linearly related to TSA, since there is no theoretical justification for a zero intercept, nor would one be anticipated using various measures of surface area. Overall, however, it is likely that the free energy of cavity formation is influenced by the size of the solute and the chemical nature of the solvent.[19]

The solute-solvent interaction term is more complex, and it is therefore more difficult to estimate its contribution. If the solute is divided into chemically different substituent groups, then a measure of $\overline{G}_{int}$ may be made by considering that each substituent type will interact with the surrounding solvent in a slightly different way. Summing all substituent interactions

$$\overline{G}_{int} = \Sigma \, k_i \, Vol_i \tag{6}$$

where k_i is a substituent dependent empirical constant with units of free energy/volume, and Vol_i is the volume of the i-th type of substituent. Division of solute volume into ΣVol_i is facilitated by recent computer programs.[19]

Combining Equations 5 and 6, the partition coefficient may be expressed by

$$RT \, \ell n \, K_{AB} = \Delta G^o_{g \to A} - \Delta G^o_{g \to B} = \Delta G^o_{B \to A}$$

$$= (k^B_{cav} - k^A_{cav}) \cdot TSA + \Sigma(k^B_i - k^A_i) \cdot Vol_i + C \tag{7}$$

where C is a constant which represents the difference in intercepts from the relationships between $\overline{G}_{cav}$ and TSA in each phase. The first term of Equation 7 may also be written in terms of molecular volume, since the use of molecular volume and has been shown to be mathematically equivalent.[23] It has been suggested[18] that the magnitude of $(k^B_i - k^A_i)$ is very small, which means that the solvent cavity formation term is the primary determinant of the extent of partitioning. If $(k^B_i - k^A_i)$ is negligible, however, then group solute-solvent interaction energies would be essentially independent of solvent type. While this may be true for similar solvents, as solvents become increasingly different, the solute-solvent interaction term is going to exert a larger influence on the observed partition coefficient.

From Equation 3, since V^B_s is approximately equal to V^A_s, $K_{AB} = \gamma_B/\gamma_A$. Volume fraction activity coefficients at infinite dilution can be expanded using Flory-Huggins theory, so that the partition coefficient K_{AB} can be expressed in terms of segments based on the various component or functional groups in solute and solvents, associated segmental volumes, and segmental solute-solvent interaction parameters. Schantz and Martire[24] used this approach

TABLE 2
Relationship Between Log K_{OW} and Log K_{hw} for Various Solute Classes

Solute class	Relationship
Alkylbenzenes	$\log K_{OW} = 0.92 \log K_{hw} + 0.15$
Alkanes	$\log K_{OW} = 0.91 \log K_{hw} + 0.01$
Alkenes	$\log K_{OW} = 1.14 \log K_{hw} - 0.93$
Bromoalkanes	$\log K_{OW} = 1.02 \log K_{hw} - 0.21$
Alcohols	$\log K_{OW} = 0.96 \log K_{hw} + 1.58$

From Schantz, M. M. and Martire, D. E., *J. Chromatogr.*, 391, 35, 1987. Copyright Elsevier Science, Amsterdam. With permission.

for the octanol/water and hexadecane/water systems to look at the effects of different solvents on the partition process. It was found theoretically that, to a good approximation, for solutes forming a homologous series

$$\log K_{OW} = 1.00 \log K_{hw} + CONSTANT$$

where K_{hw} is the hexadecane/water partition coefficient. This result forms a partial theoretical basis for Collander-type equations which relate partition coefficients in different solvent systems.[10]

To assess the effects of varying solute substituents or functional groups, values of K_{OW} and K_{hw} were experimentally determined for homologous series of alkylbenzenes, alkanes, alkenes, bromoalkanes, and normal alcohols. Linear relationships between $\log K_{ow}$ and $\log K_{hw}$ with near unit slope for all solute types confirmed the theoretical approach outlined above, and also indicated that the division of partition coefficients into contributions from various functional groups or moieties may have some fundamental theoretical basis.[10] The intercepts of the plots of $\log K_{ow}$ vs. $\log K_{hw}$ are particularly interesting since alcohols have a large positive intercept, while other solute types have essentially a zero intercept or a slightly negative one. These data are summarized in Table 2.

To interpret these results it is instructive to first consider the group of n-alkanes which consists of pentane, hexane, heptane, and octane. The near zero intercept means that K_{OW} is approximately equal in magnitude to K_{hw}. Hexadecane and octanol possess obvious dissimilarities, and it is unlikely that cavity formation energy requirements would be similar. The correspondence between K_{OW} and K_{hw} may therefore be due to a cancellation of opposing effects arising from cavity formation and solute-solvent interaction.

The homologous series of a n-alcohols exhibit a large positive intercept when $\log K_{OW}$ is plotted against $\log K_{hw}$, and therefore for a given n-alcohol K_{OW} is greater than K_{hw}. This can be rationalized by recognizing that it would be easier to form and place an alcohol molecule in a cavity in the octanol due to lower solute-solvent interaction energy; hence the free energy of solution in octanol is less, and the octanol/water partition coefficient greater than that for hexadecane/water. The series of alkylbenzenes also have a positive intercept, and this can be rationalized in a similar manner.

For alkenes and bromoalkenes the situation is reversed, in that K_{hw} is larger than K_{OW}, which means that the cavity formation and interaction processes are less demanding in the hexadecane phase. It can be seen, therefore, that although the size of the solute and the consequent size of the solvent cavity is important in determining the extent of partitioning, depending on the solute and solvent, interactions within the cavity may be important also.

B. INFLUENCE OF PHYSICAL AND CHEMICAL PROPERTIES ON PARTITIONING

1. Influence of Physical Properties of Solute

In general, it has been observed that for homologous series, or a series of closely related organic compounds, the partition coefficient increases with molecular size. As the size increases, the nonideality of the aqueous phase solution increases, which means that the aqueous solubility decreases. In the organic phase, nonideality also increases, but at a slower rate, and so the solubility of organic solutes falls off more slowly. This is described mathematically for chlorobenzenes and PCBs by the following equations:

$$\log \gamma_w^* = 1.34 + 0.0248 \ V$$

$$\log \gamma_o^* = 0.01 + 0.0048 \ V$$

where γ_w^* and γ_o^* are the solute mole fraction activity coefficients in water and octanol, respectively, and V is the solute molar volume.[19] When the solute is small in size (and hence V is small), the solutions are more ideal than those of large solutes. Since aqueous solubility decreases with size for organic compounds more quickly than octanol solubility, the ratio of solubilities (S_o/S_w) or K_{ow} increases with size for related compounds. While it is more difficult to form a cavity in both water and octanol with increasing solute size, the differences in energy requirements between water and octanol also increase due to the need to disrupt more and more solvent-solvent interactions, particularly in the aqueous phase.

There are many examples of relationships between partition coefficients and molecular descriptors of size, such as volume and TSA, in the literature. For chlorobenzenes and PCBs Miller et al.[9] found that

$$\log K_{ow} = 0.02 \ V + 0.49$$

while Leo et al.[25] obtained a similar equation for a group of chloro and alkyl substituted aromatic hydrocarbons. With regard to molecular surface area correlations, the following relationships have been found for polyaromatic hydrocarbons, alkylbenzenes, and PCB congeners, respectively:[26-28]

$$\log K_{ow} = 0.030 \ TSA - 1.389 \tag{8}$$

$$\log K_{ow} = 0.028 \ TSA - 0.863 \tag{9}$$

$$\log K_{ow} = 0.027 \ TSA - 0.996 \tag{10}$$

It is interesting to note the similarity in slopes from Equations 8 to 10. Schantz and Martire[24] have predicted that the slopes of log K_{ow} vs. molar volume plots should be the same for homologous series of solutes containing a diverse array of functional groups. This was confirmed experimentally for alkylbenzenes, alkanes, alkenes, bromoalkanes, and alcohols. Additionally, the average value of the slope for all solute types in octanol/water was not significantly different from that in hexadecane/water. Because of the correspondence between volume and surface area, the relative constancy of slope in the above equations relating log K_{ow} to TSA for different solute types is not surprising. Other measures of solute size have also been employed, such as chlorine number in the case of chlorobenzenes, and PCBs,[29] and found to result in similar linear relationships. In general, no matter what the molecular descriptor used, a direct linear relationship is obtained between the logarithm of the partition coefficient and the descriptor itself for non-ionizable, hydrophobic solutes.

TABLE 3
A Comparison of Log K_{hw}, Log K_{hepw}, Log K_{OW} and Log K_{tw} Values for Alkylbenzenes

	log K_{tw}	log K_{hepw}	log K_{OW}	log K_{tw}
Benzene	2.02	2.26	2.13	2.25
Toluene	2.71	2.85	2.69	2.77
Ethylbenzene	3.23	3.43	3.15	3.27
n-Propylbenzene	3.84	4.11	3.68	3.77

Compiled from Chiou and Block[15] and Schantz and Martire.[24]

2. Influence of Chemical Properties of Solute

The extent and importance of the influence of chemical properties of the solute on the partition coefficient is best illustrated by an example. The molecular volumes of n-hexane and n-hexanol are roughly equal, but the K_{OW} values for these two compounds are $10^{4\,16}$ and $10^{2\,05}$, respectively.[24] As discussed previously, the slopes of plots of log K_{OW} vs. molar volume for homologous series of n-alkanes and n-alcohols are the same, but obviously the intercepts are different. Given the difference in partition coefficient, due primarily to the presence of different functional groups, log K_{OW} values of both solute types respond in a similar manner to the successive addition of methylene (CH_2) fragments. Because the free energies of cavity formation for n-hexane and n-hexanol are likely to be roughly the same, the partition coefficient difference is a reflection of different solvent-solute interaction energies.

For a homologous series of non-ionic surfactants, Cratin[30] suggested that the free energy of transfer, or partitioning, could be divided up into contributions from "n" hydrophilic groups and a hydrophobic group. Analogously, for hydrophobic organic molecules, a similar subdivision into group fragments should also be possible. On plotting log K_{OW} vs. n, which is a measure of size, the intercept is related to the free energy of transfer for the functional group. Schantz and Martire[24] have determined the free energy of transfer of the hydroxyl function to be 9.71 kJ mol.$^{-1}$ The positive value implies a preference for the aqueous phase and indicates that log K_{OW} for n-hexanol is less than the value for n-hexane largely because of decreased solute-solvent interaction free energy in the aqueous phase, and increased interaction free energy in the organic phase.

The extent of partitioning is affected not only by the nature of any functional groups present, but also by the number and position of substitution of such groups. As an example, the log K_{OW} value of 3,5-dinitrophenol is 2.32, but the value for the isomeric 2,6-dinitrophenol is over an order of magnitude less at 1.18, due to the closer proximity of the nitro groups to the hydroxyl functionality.[10]

3. Influence of Physical Properties of Solvent

The physical property likely to be of most importance is the size of the solvent. Intermolecular attractions generally increase with size, as evidenced by increasing boiling points within homologous series. For a given solute, to form a cavity in a relatively large solvent would necessitate disrupting stronger intermolecular solvent-solvent interactions than cavity formation in a smaller solvent. On this basis, free energies of solution in hexadecane, for instance, would be greater than those required for solution in heptane, and so $K_{hw} < K_{hepw}$, where K_{hepw} is the heptane/water partition coefficient. To test this hypothesis, the K_{hw} and K_{hepw} values for a series of alkylbenzenes are compiled in Table 3. It is apparent that K_{hw} is consistently less than K_{hepw}, which is compatible with initial predictions. Since $K_{hw} < K_{hepw}$, cavity formation for solutes such as alkylbenzenes is easier and less energetically demanding in heptane.

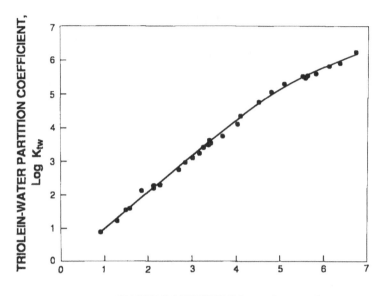

FIGURE 2.   A comparison of log K_{tw} vs. log K_{ow} for some organic compounds. (From Chiou, C. T., *Environ. Sci. Technol.*, 19, 57, 1985. Copyright American Chemical Society. With permission.)

To determine whether this effect of slightly decreasing partition coefficient with increasing organic solvent size is also applicable to more polar solvents, Table 3 also contains partition coefficients of alkylbenzenes for the octanol/water and triolein/water (K_{tw}) systems. Here, the magnitude of the partition coefficient of the larger solvent is slightly greater than that of the smaller solvent ($K_{tw} > K_{OW}$).

Chiou[3] has compared K_{tw} with K_{OW} for a wider variety of solutes, ranging from aniline and anisole to hexachlorobiphenyl. The plot of this correlation is found in Figure 2. Its striking feature is the nonlinearity for larger, more hydrophobic solutes. With compounds possessing log $K_{OW} < 5$, the linear regression between log K_{tw} and log K_{OW} is given by

$$\log K_{tw} = 1.00 \log K_{OW} + 0.11$$

which is the result expected on the basis of the alkylbenzene data in Table 3.

It has been proposed that the reason for the curvilinearity observed in Figure 2 is the large size disparity between octanol and triolein. The molar volume of octanol is approximately 130 cm³ mol,$^{-1}$ while that of triolein is some 7 times larger.[3] Much of the preceding theoretical discussion concerning partitioning assumes regular solution theory and a Raoult's Law convention for activity coefficients. The effect of solvent size may be accounted for by considering partitioning in terms of an adaption of the Flory-Huggins theory, which deals with solution in macromolecular phases.[31] Using this approach, log K_{tw} may be expressed by

$$\log K_{tw} = \log (\gamma_w^* V_w) - \log V_s^t - \left(1 - \frac{V_s^t}{V_t}\right)/2.303 - \chi/2.303 \qquad (11)$$

where V_w and V_t are the molar volumes of water and triolein, respectively, V_s^t the solute molar volume in triolein, and χ the Flory-Huggins interaction parameter, which is a measure of solute-solvent interaction. It should be noted that if $V'_s \approx V_1$, and $\chi/2.303$ approximates the activity coefficient in the organic phase, then Equation 11 is equivalent to Equation 4, which has been used for K_{OW}, and describes partitioning on a Raoult's law basis.[3] Equation 11 can therefore be considered to be a more complete treatment of partitioning, encompassing solute-solvent size disparity if necessary.

By employing expressions of the form of Equation 11, we can now assess the effects of solvent size on partition coefficients. The third term of such an expression, i.e., $(1 - V'_s/V_1)/2.303$ is larger for triolein than for octanol, and this would therefore tend to make $K_{tw} < K_{OW}$. For solute molecules whose molar volumes are roughly that of octanol, the incompatibility term is greater for octanol than for triolein, which tends to make $K_{OW} < K_{tw}$. With such solutes, incompatibility is negligible in triolein because of the difference in solute and solvent size. Since there is only a slight difference found experimentally between K_{OW} and K_{tw} for solutes of this size (0.1 log units), the two opposing tendencies described above must to a large extent cancel. Only for relatively large solutes will the incompatibility term have an effect for triolein. With such compounds, the third term of Equation 11 may dominate for both triolein and octanol, meaning that K_{tw} would become less than K_{OW}.

In summary, for relatively small solutes (i.e., those whose molar volumes are less than around 250 cm^3 mol^{-1}) the partition coefficient for nonpolar organic solvents will be smaller with increasing solvent size. The reverse is the case for more polar solvents and small solutes with slightly increasing partition coefficients accompanying increased solvent size. However, larger solutes eventually result in decreasing partition coefficients with increasing solvent size.

4. Influence of Chemical Properties of Solvent

According to the cavity concept of partitioning, the free energy required to create a cavity depends upon its size, which is related to the size of the solute. Once the solute is placed within the cavity, the extent to which it interacts with the solvent has been shown to depend upon solute chemical properties.[23] Solvent chemical properties must also play a role, and the property most likely to do so is polarity, as determined by the nature and number of functional groups contained by the solvent.

The influence of solvent polarity is most conveniently investigated by considering partition coefficients in solvent systems where the organic phases are approximately similar in molecular size, but differ in polarity. These criteria are met by the octanol/water and heptane/water systems. A plot of log K_{OW} vs. log K_{hepw} for a variety of solutes based on data compiled by Chiou and Block[16] is found in Figure 3. For smaller, more hydrophilic solutes, such as aniline, acetophenone, benzoic acid, and nitrobenzene, log K_{OW} is greater than log K_{hepw}. This is denoted by the fact that the points representing these compounds lie above the dashed line which indicates $K_{OW} = K_{hepw}$. Since the polarities of solute and solvent are more similar in octanol, free energy of solution is less in octanol compared with heptane. Therefore, K_{OW} is greater than K_{hepw}.

The larger more hydrophobic solutes such as PCBs are found below the dashed line in Figure 3, which means that in these cases log K_{hepw} is greater than log K_{OW}. The reason for this is analogous to that presented above. Polarities are now more similar between heptane and solute, so the free energy of solution is less in heptane, and K_{hepw} is greater than K_{OW} Linear regression of log K_{OW} against log K_{hepw} results in

$$\log K_{OW} = 0.71 \log K_{hepw} + 1.02 \qquad (r = 0.95, \quad n = 31)$$

which is represented by the solid line in Figure 3.

Similar trends are exhibited by data experimentally determined by Schantz and Martire.[24] For example, with 1-butanol, its log K_{OW} value of 0.79 is larger than its log K_{hw} value of -0.96, while for n-non-1-ene, log K_{OW} (5.14) is less than log K_{hw} (5.36).

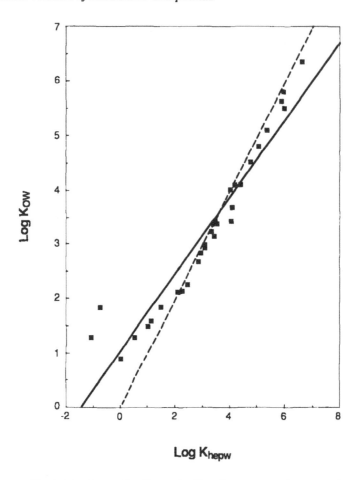

FIGURE 3.   A plot of log K_{OW} vs. log K_{hepw} for some aromatic compounds, with the continuous line representing the linear regression equation, and the dashed line representing log K_{OW} = log K_{hepw}.

III. IMPORTANT MOLECULAR CHARACTERISTICS IN PARTITIONING

A. SURFACE AREA

1. Definitions

The previous section has highlighted the roles of solute size and solvent cavity formation in the partitioning process, and the importance of molecular descriptors such as volume and TSA in predicting partition coefficients. Pearlman[23] has stated that since the molecules consist of vibrating nuclei surrounded by a diffuse cloud of electrons, any definition of molecular surface, and therefore surface area, is inherently artificial. Nevertheless, if the van der Waals radius of an atom can be defined as the distance of closest approach between such atoms that are not bonded, then each atom has a spherical van der Waals surface area based upon this radius.[32] The van der Waals surface area of a molecule is then defined as the surface of intersection of all atomic surface areas in the molecule.

Since not all of the van der Waals surface area may be accessible to solvent molecules, the solvent accessible surface area has been defined as that surface area generated or traced by the center of a solvent sphere that is moved over the van der Waals surface area of a solute. It should be noted that the solvent accessible surface area is equivalent to the van der Waals surface area if the solvent radius is zero.

The solvent accessible surface has little physical meaning, however; therefore, the contact surface area has been defined as the area traced by the edge of a solvent sphere that is moved over the van der Walls surface area. This surface area excludes those parts of the solute that are inaccessible to the solvent for steric reasons, and can be considered as an indication of the extent to which the solute is exposed to intermolecular interaction with the solvent.[23]

2. Measurement and Estimation

One of the earliest methods for measuring molecular surface area was the simple summation of the van der Waals surface areas of the component atoms of the molecule.[33] As an extension of this, surface areas have also been calculated by considering a molecule as a collection of spherical groups of atoms, for example OH and CH_2, including a solvent of arbitrary radius.[34]

Physical methods of surface area estimation have included the covering of molecular models with styrofoam balls which represent solvent molecules. The number of spheres able to be attached give a measure of the surface area. Such a procedure can only result in relative surface areas, with the t-butyl group often assigned a value of one.[21,35] However, a crude estimate of the accessible surface area reflecting solvent size is obtained.

Recently, a number of computer programs have become available that calculate surface area very precisely. Overlap between constituent atom surface areas is accounted for, and Equations such as 8,9 and 10, where surface area is related to partition coefficients, generally are derived using computer based algorithms. One program often employed is that written by Hermann,[36] in which van der Waals and accessible surface areas can be calculated. Most recently, Pearlman[23] has successfully devised a program that enables the contact surface area to be obtained with any desired precision.

B. VOLUME
1. Definitions

Just as TSA or surface area can be defined and measured by a variety of methods, volume can be defined in a number of ways, each resulting in a slightly different estimate. On a macroscopic scale, molar volume is defined as the volume occupied by one mole of the compound of interest. This measure of volume is not simply molecular volume multiplied by Avogadro's number, as it encompasses intermolecular volume as well. The van der Waals, solvent accessible, and contact surface areas introduced previously all have specific volumes associated with them. Another volume measure is the parachor of a compound. This parameter, related to surface tension and molecular weight, together with liquid and vapor densities, has been described as a surface tension corrected molar volume.[37,38]

2. Measurement and Estimation

Molar volume can be either measured directly, or estimated. One estimation method useful for hydrocarbons and related compounds is to count the number of carbon, hydrogen, oxygen, and nitrogen atoms, add one for each double bond, then multiply the sum by seven.[39] A similar widely used procedure is the Le Bas method, in which molar volume is calculated by summing constituent atom volumes, then correcting for unsaturation and various ring structures. While affording a reasonable estimate of volume, such procedures have been found to be inadequate in correlating observed partition coefficient variation with subtle volume changes for solutes such as chlorobenzene and PCB congeners.[40] Parachors may also be calculated by additive constituent constants and subsequent correction formulae.[38]

Leo et al.[25] estimated molecular volume by wrapping Corey-Pauling-Koltun models in stretchable polyethylene, sealing the assembly with tape, and carefully measuring the amount of water displaced. Results obtained using this method were comparable to those of Bondi[33] calculated from van der Waals volumes for a wide range of aliphatic and aromatic hydro-carbons.

The computation of volumes is a far more intricate and complex procedure than that for surface areas. Much effort is currently being devoted to this since the utility of molecular descriptors in describing the partitioning, and to some extent the environmental behavior of persistent hydrophobic chemicals, has been realized. Algorithms for the computation of van der Waals and solvent accessible volumes have appeared, and programs for the calculation of contact volume are currently being developed.[23]

IV. PARTITIONING IN ORGANISM/WATER SYSTEMS

A. SIMILARITIES AND DIFFERENCES WITH ABIOTIC PARTITIONING SYSTEMS

1. Similarities

Aquatic organisms are very complex when compared with simple abiotic two-phase partitioning systems. But similarities between K_B and partition coefficients of the same compounds in various abiotic systems have prompted suggestions that dialysis membranes filled with lipophilic solvent could serve as a monitor or indicator of contamination by xenobiotic compounds, especially in environments too severe for biological indicators to exist. Analytical procedures would be greatly simplified using this approach as well. Söd-ergren has found that under both laboratory and field conditions, solvent-filled dialysis bags concentrated organochlorines such as p,p'-DDT, and PCB congeners, in a qualitatively similar manner to various species of aquatic biota, including the amphipod *Gammarus pulex*.[41]

The uptake and depuration rates, together with times to equilibrium for abiotic concentration, are dependent upon the dialysis membrane surface area and membrane permeability, and bear little relationship with analogous parameters from bioconcentration. However, by appropriate choice of solvent, the observed solvent/water partition coefficient may be equivalent to the bioconcentration factor on a lipid basis for many persistent compounds.

2. Differences

Xenobiotic chemicals likely to be bioconcentrated vary widely in their size and hydro-phobicity, as measured by K_{OW}. Equivalence between K_B on a lipid basis and K_{OW}, or linear relationships between log K_B and log K_{OW}[42] have been found only for nondegradable compounds with log K_{OW} values between 2 and 6. Below the limiting log K_{OW} value of 2, chemicals bioconcentrate more than would be anticipated on the basis of these expressions. Conversely, above log $K_{OW} = 6$, chemicals tend to bioconcentrate less than expected.

In the case of hydrophilic compounds (log $K_{OW} < 2$), the amount of solute in nonlipid fractions of the organism may become appreciable,[42] and therefore bioconcentration can no longer be treated simply as a partitioning between the surrounding water and lipid tissue. Gobas et al.[1] have suggested that the loss of linearity between log K_B and log K_{OW} observed for extremely hydrophobic (log $K_{OW} > 6$) compounds is due to factors affecting the linear relationship between the solute's activity coefficients in the lipid phase and the octanol phase. A contributing factor may be the gross size disparity between lipid and octanol. Earlier in this chapter, it was suggested that the size difference between triolein and octanol was a primary reason for the curvilinearity noticeable in the plot of log K_{tw} vs. log K_{OW}, depicted in Figure 2. Since lipid is as large or even slightly larger than triolein, then the negative deviation from linearity on plotting log K_B vs. log K_{OW} would be expected to be as large, or possibly larger, than that in Figure 2. The reason is analogous, in that for larger solutes decreasing solubility and increasing nonideality in lipid means that log $K_B <$ log K_{OW}.

The possible significance of lipid solubility in bioconcentration has been investigated

TABLE 4

Free Energy, Enthalpy, and Entropy Changes (kJ Mol^{-1}) During Transfer of Chlorobenzenes from Water to Fish and to Octanol at 292K

Solute	log K_{OW}	$\Delta G°$	$\Delta H°$	$T\Delta S°$	log K_B	$\Delta G°$	$\Delta H°$	$T\Delta S°$
1,3 Dichlorobenzene	3.55	−25.0	−15.0	10.0	3.78	−26.3	6.0	32.3
1,3,5-Trichlorobenzene	4.32	−29.4	−21.7	7.7	4.35	−29.5	10.5	40.0
1,2,3,4-Tetrachlorobenzene	4.61	−31.0	−26.3	4.7	4.74	−31.7	13.5	45.2
Pentachlorobenzene	5.05	−33.6	−30.8	2.8	5.20	−34.2	16.0	50.2
Hexachlorobenzene	5.70	—	—	—	5.66	−36.6	17.9	54.3

From Opperhuizen, A., Serne, P. and Van der Steen, J. M. D., *Environ. Sci. Technol.*, 22, 286, 1988. Copyright American Chemical Society. With permission.

by Dobbs and Williams,[43] who found decreasing lipid solubilities with increasing solute hydrophobicity. Certainly for many large chemicals, such as polychlorinated naphthalenes,[44] dibenzodioxins,[45] perchlorobiphenyl, and perchloro-p-terphenyl,[46] bioconcentration factors are less than anticipated on the basis of hydrophobicity, as measured by K_{OW}. This leads to speculation as to whether octanol is the most appropriate solvent for use in partition coefficient-K_B relationships. Ho et al.[47] found that n-butanol/water partition coefficients were in fact a better predictor of lipid/water partition coefficients than K_{OW}.

Recently, it has been suggested that any correspondence between K_B on a lipid basis and K_{OW} is purely coincidental, and the result of a balancing of enthalpic and entropic effects. The Gibbs free energy of transfer between two phases can be divided into enthalpic ($\Delta H°$) and entropic ($\Delta S°$) contributions by

$$RT \ln K_{AB} = -\Delta G°_{A \to B} = -(\Delta H°_{A \to B} - T\Delta S°_{A \to B})$$

By measuring fish/water and octanol/water partition coefficients at selected temperatures, enthalpies and entropies of transfer for the transport systems are calculable. This has been done for some chlorobenzene congeners and the results summarized in Table 4. This indicates that for bioconcentration, water to fish transfer is enthalpically unfavorable because of the structured nature of lipids. The process is actually driven by a very favorable entropy change. For octanol/water partitioning on the other hand, the enthalpy is favorable, perhaps due in this case to interaction between the hydroxyl group of octanol and the aromatic ring structure of the solute. The entropy contribution is only weakly favorable, showing that in both water and octanol an increase in the solvent structuring occurs after solution. Thus, any loss of correspondence between K_B and K_{OW} may simply be a result of loss of enthalpy-entropy compensation in one or both systems.[48]

B. ROLE OF MEMBRANES

One outstanding difference between biotic and abiotic partitioning systems is the presence of a membrane with organisms. Although the composition of membranes varies according to their origin, they generally comprise approximately 40% of their dry weight as lipid and 60% as protein, held together by hydrogen bonding and other nonbonded interactions.[49] Modern views of membrane structure are that it consists of lipid material organized into a bilayer, embedded in which are the receptors, enzymes, and specific transport systems making up the protein constituents.[50]

In terms of passage through membranes by hydrophobic molecules unaffected by specific transport mechanisms, the structure can be thought of as consisting of five regions arranged in series. On proceeding through the membrane, a molecule encounters successively (1) an

TABLE 5
Free Energies of Transfer (kJ Mol^{-1}) of Various Functional Groups from Water to Organic Solvents and the Membrane of the Alga *Nitella* sp.

Group	Solvent					
	i-Butanol	Octanol	Ether	Olive oil	Alkane	Nitella membrane
–OH	4.18	6.61	8.79	11.7	22.1	15.1
–O–	2.51	5.61	5.86	5.86	—	3.35
–C=O	—	6.90	8.79	9.20	10.4	10.50
–COOH	5.02	—	5.02	5.86	—	5.86
–CONH$_2$	7.11	9.75	20.5	20.1	26.9	25.9
–NHCONH$_2$	7.95	—	23.0	22.2	36.5	30.5
–CH$_2$–	−2.22	−2.85	−2.80	−2.76	−3.81	−2.55

From Anderson, B. D., *Physical Chemical Properties of Drugs*, Vol. 10, Chap. 7, 1980. Copyright Marcel Dekker. With permission.

outer region consisting of the polar portion of the lipids that is relatively flexible, (2) a less polar, tightly packed area consisting of sterol ring systems, the glycerol moiety, and the first few hydrocarbon groupings, and (3) a nonpolar, flexible region comprising the hydrocarbon chains of the lipid materials followed by two regions very similar in composition to (2) and (1), respectively, but differing slightly, reflecting the asymmetry of most membranes. Basic to the function of the membranes is the ability to select or discriminate between molecules.[51] Stein[50] has proposed that this is achieved by virtue of their acting as an organic solvent, and discriminating according to partition behavior, and also by membranes acting as structured soft polymers, or very viscous fluids, and discriminating according to solute size.[52]

Since traversal may involve factors other than partitioning, membrane/water partition coefficients may bear little relationship with organic solvent/water partition coefficients. In support of this, Ho et al.[47] has shown that only a poor correlation exists between permeability and K_{OW} for a variety of solutes, including alcohols, steriods and pyridine derivatives. In addition, membranes are known to discriminate against branched hydrocarbons more strongly than would be expected on the basis of their K_{OW} values.[34] This has been attributed to the fact that lipid molecules in membranes possess a more highly ordered structure than bulk solvents, so that the lipid structure may operate like a molecular sieve in preferentially retarding branched solutes.[50]

To place the difference between organic solvents and membranes on a quantitative basis, Table 5 contains free energies of transfer for various functional groups from water to organic solvents and the membrane of the alga *Nitella* sp.[53] The data measure the amount by which each group changes the difference between a solute's partial molar free energy of solution in water, and its partial molar free energy of solution in an organic solvent, or the *Nitella* membrane.[51] It is clear that the methylene group contribution is relatively insensitive to the nature of the partitioning system. This is not the case for more polar functional groups such as the hydroxyl, whose influence varies markedly, depending on the solvent employed. Furthermore, it is evident that no single solvent can adequately model the membrane/water partitioning process for solutes containing a variety of functional groups.

The reduced accumulation of extremely hydrophobic chemicals, such as perchloronaphthalene, perchlorodibenzo-p-dioxin, and perbromobenzene in aquatic organisms, may indicate that thermodynamic equilibrium across the membrane is prevented.[44] It has been suggested that there is a molecular weight cutoff of approximately 500 to 600 Dalton for large solutes.[54] However, in one investigation perchlorodibenzo-p-dioxin (mol wt 460) was not accumulated to any extent, while decachlorobiphenyl (mol wt 499) was observed to exhibit a small linear accumulation with time.[45] Therefore, molecular weight may not be the most appropriate parameter for describing membrane permeation.

For fish, the most important membrane surfaces for bioconcentration are the gills. Gill uptake efficiencies for xenobiotics have a parabolic relationship with log K_{OW}.[55] For compounds with log K_{OW} values between 3 and 6, the uptake efficiency has been found to be approximately 60%. This may be compared with the measured O_2 uptake efficiency of 62%. Such compounds thus appear to be efficiently uptaken through the gills; they are also bioconcentrated to an extent consistent with simple partitioning, and predictable from K_{OW}. Larger compounds, with log K_{OW} values greater than 6, have considerably lower uptake efficiencies. Mirex, for example (log K_{OW} = 7.50), has an extraction efficiency of only 20%, and on this basis McKim[55] has suggested that chemicals with log K_{OW} greater than 8 may not be significantly accumulated across the gills of fish, perhaps affecting the observed bioconcentration factor.

The reason for differential extraction efficiency of xenobiotics may be found by treating membranes as soft polymers. Diffusion through soft polymers is considered to occur via transient regions of free volume or holes, created by random thermal motion of the polymer chains.[52] In terms of lipid membranes, passage of hydrophobic molecules can occur only when cavities are formed on the polar outer surface by random motion of the lipids. If the constitution of membrane is constant over time, there is likely to be a maximum size of the cavities created.[44] This size will in turn determine the maximum size of permeable chemicals. For permeation to occur, two of the three dimensions of the permeating molecules must be less than the smallest cross section of the cavity. From work with chlorinated naphthalenes, Opperhuizen et al.[44] have proposed an absence of membrane permeation for hydrophobic molecules with effective cross sectional areas greater than $9.5 \times 10^{-10} m^2$.

It is unlikely that membranes affect the final equilibrium position between biotic lipid and water due to their relatively small volume. They may however perturb the *observed* bioconcentration factor for large, extremely hydrophobic solutes from that expected by significantly reducing uptake rates, and effectively isolating storage lipid compartments, so that equilibrium is never achieved within a practical timespan.

C. MODELS FOR BIOTA/WATER PARTITIONING

Since few active processes are known that facilitate traversal of hydrophobic compounds across membranes, they are generally assumed to cross by simple, passive diffusion mechanisms.[55] Once inside the organism, they need to be transported to various lipid repositories. Studies on drug transport have shown that this process can often be considered as passive and diffusion controlled,[56] although real fluid flow is important in some instances as well. It could be expected that hydrophobic molecules are influenced by the same factors; therefore, it is not surprising that many models of bioconcentration and lipid/water partitioning are based on diffusion principles.

One such model proposed by Gobas et al.[1] describes a fish as an emulsion of lipid in a volume of water, separated from the ambient water by membranes. Clearly this model does not account for internal distribution processes. Using such an approach, uptake and depuration rate constants (k_1 and k_2, respectively) have been shown to be controlled either by permeation through the membrane, or by permeation through unspecified aqueous phases. For smaller, more hydrophilic solutes, the controlling factor is membrane permeation, while for larger, more hydrophobic solutes, the latter factor is dominant.

If the ambient water concentration remains constant, then uptake under membrane permeation control may be expressed by

$$k_1 = \frac{K_m D_m A}{\delta_m F} \tag{12}$$

where K_m is the membrane/water partition coefficient, D_m is the solute diffusion coefficient in the membrane, δ_m is the membrane thickness, A is the diffusion area (which is effectively the membrane surface area), and F is the weight of the fish. Clearance under similar conditions is described by

$$k_2 = \frac{D_m A}{\delta_m F} \left(\frac{1 - \ell}{K_m} + \ell \right)^{-1} \tag{13}$$

where ℓ is the ratio of the total lipid weight to the fish weight. The derivation of Equation 13 assumes that the lipid/water partition coefficient is equal to the membrane/water partition coefficient. The fish can be considered as a single inhomogeneous compartment, and at equilibrium $K_B = k_1/k_2$. Therfore, in terms of Equations 12 and 13

$$K_B = \ell K_m + (1 - \ell) \tag{14}$$

Since hydrophobic compounds in bioconcentration experiments usually show that ℓK_m is much greater than $1 - \ell$,[1] Equation 14 can be approximated and rewritten as

$$\log K_B = \log K_m + \log \ell \tag{15}$$

Analogous deviation of k_1 and k_2 for xenobiotic solutes under aqueous phase diffusion layer control also yields Equation 15. If membrane and repository lipids are adequately simulated by octanol, then this equation assumes the traditional form of K_B vs. K_{OW} relationships. This result demonstrates that factors related to membrane permeation cannot affect a chemical's final equilibrium partitioning position, and hence ultimate K_B value. Therefore, as mentioned previously, any loss of linearity between $\log K_B$ and $\log K_{OW}$ is probably fundamentally due to differences arising in activity coefficients in the lipid and octanol phases,[1] loss of compensation between enthalpic and entropic contributions toward partitioning,[48] or perhaps insufficient exposure time to attain equilibrium.

Another approach to modeling biota/water partitioning is to treat the organism, a fish for example, as a multicompartment system in which all compartments have unique individual kinetic expressions for uptake and clearance. In order to reduce the complexity of this model, it is first assumed that a steady state situation applies to all compartments except the one containing the storage lipid phase. Second, it is assumed that all internal compartments are either organic or aqueous in nature.[57] Because we are concerned now only with the surrounding aqueous medium and the lipid phase, the bioconcentration factor derived will be on a lipid basis. At equilibrium, the fugacities of these aqueous and lipid phases will be equal, and since for a given phase $f \approx C/Z$ where C is the solute concentration and Z a fugacity capacity constant,[17,55]

$$K_B = Z_L/Z_W$$

Clearly, if Z_L, the lipid fugacity capacity constant equals Z_o, then K_B on a lipid basis can be equated with K_{OW}. Again, however, if there is a difference in the activity coefficients of lipid and octanol phases for a solute, then it might be expected that Z_L would differ from Z_o, and hence the linear relationship between K_B and K_{OW} would break down.

Bruggeman et al.[58] have developed a model for bioconcentration by fish based upon partitioning between phases in which phases represent actual anatomical units. The storage compartment was the body lipid phase, while the blood serves as a transport compartment which has a low storage capacity, but a high turnover rate. Chemicals enter through a so-called gate compartment, where the most efficient interphase transfer occurs, and this rep-

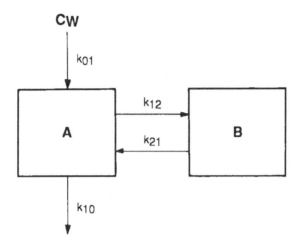

FIGURE 4. A two compartment organism model with uptake and elimination from only one compartment. (From Spacie, A. and Hamelink, J. L., *Environ. Toxicol. Chem.*, 1, 309, 1982. Copyright Pergamon Press, Elmsford, NY. With permission.)

resents the gill structure. In this model, the gate compartment is considered to be largely aqueous in nature. For uptake, the rate of transport from the surrounding water to the storage phase is equal to the activity gradient divided by the sum of intervening resistance factors. Clearance is not as straightforward a process, however, since a two-phase clearance is possible. This will occur if the transfer rate constant between lipid and blood is of a similar order of magnitude to that between the blood and gill compartment.[58] At equilibrium, the bioconcentration factor on a lipid basis is given by the ratio of the solute activity coefficient in lipid and ambient water phases.

Biphasic clearance behavior is also observed with models such as that illustrated schematically in Figure 4. Initial rapid elimination represents clearance from the peripheral compartment A, while the slower second step is the result of redistribution from a second compartment within the organism.[5] Clearly, with two internal compartments capable of concentration, partitioning in general, and K_B in particular, can no longer be interpreted in terms of partitioning between ambient water and a single lipid compartment. In such cases, K_B can only be described by large and complex expressions involving the rate constants depicted in Figure 4.[59] Other studies have employed even more detailed models, involving for example, three internal lipid compartments.[60]

At present, biodegradation is a complicating factor, and kinetic models based on diffusion are useful as tools for understanding the process of bioconcentration only in the absence of biodegradation. Since most xenobiotics are degradable to some extent, a more complete understanding of bioconcentration in terms of partitioning may perhaps be achieved by adapting some analogous pharamaco-kinetic models.

REFERENCES

1. **Gobas, F. A. P. C., Opperhuizen, A., and Hutzinger, O.,** Bioconcentration of hydrophobic chemicals in fish: Relationship with membrane permeation, *Environ. Toxicol. Chem.*, 5, 637, 1986.
2. **Davies, R. P. and Dobbs, A. J.,** The prediction of bioconcentration in fish, *Water Res.*, 18, 253, 1984.
3. **Chiou, C. T.,** Partition coefficients of organic compounds in lipidwater systems and correlations with fish bioconcentration factors, *Environ. Sci. and Technol.*, 19, 57, 1985.

4. **Geyer, H., Scheunert, I., and Korte, F.,** Relationship between the lipid content of fish and their bioconcentration potential of 1,2,4-trichlorobenzene, *Chemosphere,* 14, 545, 1985.

5. **Konemann, H. and Van Leeuwen, K.,** Toxicokinetics in fish: Accumulation and elimination of six chlorobenzenes by guppies, *Chemosphere,* 9, 3, 1980.

6. **Zaroogian, G. E., Heltshe, J. F., and Johnson, M.,** Estimation of bioconcentration in marine species using structure-activity models, *Environ. Toxicol. Chem.,* 4, 3, 1985.

7. **Watarai, H., Tanaka, M., and Suzuki, N.,** Determination of partition coefficients of halobenzenes in heptane/water and 1-octanol/water systems and comparison with the scaled particle calculation, *Anal. Chem.,* 54, 702, 1982.

8. **Briggs, G. G.,** Theoretical and experimental relationships between soil adsorption, octanol-water partition coefficients, water solubilities, bioconcentration factors and the parachor, *J. Agric. Food Chem.,* 29, 1050, 1981.

9. **Miller, M. M., Wasik, S. P., Huang, G. -L., Shiu, W. -Y., and Mackay, D.,** Relationships between octanol-water partition coefficient and aqueous solubility, *Environ. Sci. Technol.,* 19(6), 522, 1985.

10. **Leo, A., Hansch, C., and Elkins, D.,** Partition coefficients and their uses, *Chem. Rev.,* 71, 525, 1971.

11. **Maron, S. H. and Lando, J. B.,** *Fundamentals of Physical Chemistry,* Macmillan, New York, 1974, 466.

12. **Chiou, C. T., Freed, V. H., Schmedding, D. W., and Kohnert, R. L.,** Partition coefficient and bioaccumulation of selected organic chemicals, *Environ. Sci. Technol.,* 11, 475, 1975.

13. **Laidler, K. J. and Meiser, J. H.,** *Physical Chemistry,* Benjamin/Cummings, Menlo Park, California, 1982, Chap. 5.

14. **Tewari, Y. B., Martire, D. E., Wasik, S. P., and Miller, M. M.,** Aqueous solubilities and octanol-water partition coefficients of binary liquid mixtures of organic compounds at 25°C. *J. Solution Chem.,* 11, 435, 1982.

15. **Chiou, C. T. and Block, J. H.,** *Partition Coefficient—Determination and Estimation,* Pergamon Press, Elmsford, NY, 1986, Chap. 3.

16. **Lewis, G. N. and Randall, M.,** *Thermodynamics,* 2nd Ed., McGraw-Hill, New York, 1961, Chap. 14.

17. **Mackay, D. and Paterson, S.,** Calculating fugacity, *Environ. Sci. Technol.,* 15, 1006, 1981.

18. **Pearlman, R. S.,** *Partition Coefficient—Determination and Estimation,* Pergamon Press, Elmsford, NY, 1986, Chap. 1.

19. **Pierotti, R. A.,** A scaled particle theory of aqueous and nonaqueous solutions, *Chem. Rev.,* 76, 717, 1976.

20. **Kamlet, M. J., Doherty, R. M., Carr, P. W., Mackay, D., Abraham, M. H., and Taft, R. W.,** Linear solvation energy relationships. 44. Parameter estimation rules that allow accurate prediction of octanol/water partition coefficients and other solubility and toxicity properties of polychlorinated biphenyls and polycyclic aromatic hydrocarbons, *Environ. Sci. Technol.,* 22, 503, 1988.

21. **Reynolds, J. A., Gilbert, D. B., and Tanford, C.,** Empirical correlation between hydrophobic free energy and aqueous cavity surface area, *Proc. Natl. Acad. Sci. U.S.A.,* 71, 2925, 1974.

22. **Hermann, R. B.,** Use of solvent cavity area and number of packed solvent molecules around a solute in regard to hydrocarbon solubilities and hydrophobic interactions. *Proc. Natl. Acad. Sci. U.S.A.,* 74, 4144, 1977.

23. **Pearlman, R. S.,** Molecular surface areas and volumes and their use in structure/activity relationships, in *Physical Chemical Properties of Drugs,* Yalkowsky, S. H., Sinkula, A. A., and Valvani, S. C., Eds., Marcel Dekker, New york, 1980, Chap. 10.

24. **Schantz, M. M. and Martire, D. E.,** Determination of hydrocarbon-water partition coefficients from chromatographic data and based on solution thermodynamics and theory, *J. Chromatogr.,* 391, 35, 1987.

25. **Leo, A., Hansch, C., and Jow, P. Y. C.,** Dependence of hydrophobicity of apolar molecules on their molecular volume, *J. Med. Chem.,* 19, 611, 1976.

26. **Yalkowsky, S. H. and Valvani, S. C.,** Solubilities and partitioning. 2. Relationships between aqueous solubilities, partition coefficients and molecular surface areas of rigid aromatic hydrocarbons, *J. Chem. Eng. Data,* 24, 127, 1979.

27. **Yalkowsky, S. H. and Valvani, S. C.,** Partition coefficients and surface areas of some alkylbenzenes, *J. Med. Chem.,* 19, 727, 1976.

28. **Burkhard, L. P., Armstrong, D. E., and Andren, A. W.,** Partitioning behavior of polychlorinated biphenyls, *Chemosphere,* 14, 1703, 1985.

29. **Miller, M. M., Ghodbane, S., Wasik, S. P., Tewari, Y. B., and Martire, D. E.,** Aqueous solubilities, octanol/water partition coefficients and entropies of melting of chlorinated benzenes and biphenyls, *J. Chem. Eng. Data,* 29, 184, 1984.

30. **Cratin, P. D.,** Partitioning at the liquid-liquid interface, *Ind. Eng. Chem.,* 60, 14, 1968.

31. **Chiou, C. T., Porter, P. E., and Schmedding, D. W.,** Partition equilibria of nonionic organic compounds between soil organic matter and water, *Environ. Sci. Technol.,* 17, 227, 1983.

32. **Laidler, K. J. and Meiser, J. H.,** *Physical Chemistry,* Benjamin/Cummings, Menlo Park, California, 1982, Chap. 15.

33. **Bondi, A.,** Van der Waals volumes and radii, *J. Phys. Chem.,* 68, 441, 1964.

34. **Valvani, S. C., Yalkowsky, S. H., and Amidon, G. L.,** Solubility of nonelectrolytes in polar solvents. VI. Refinements in molecular surface area computations, *J. Phys. Chem.,* 80, 829, 1976.

35. **Harris, M. J., Higuchi, T., and Rytting, J. H.,** Thermodynamic group contributions from ion pair extraction equilibria for use in the prediction of partition coefficients. Correlation of surface area with group contributions, *J. Phys. Chem.,* 77, 2694, 1973.

36. **Hermann, R. B.,** Theory of hydrophobic bonding. II. The correlation of hydrocarbon solubility in water with solvent cavity surface area, *J. Phys. Chem.,* 76, 2754, 1972.

37. **Oliver, B. G. and Niimi, A. J.,** Bioconcentration factors of some halogenated organics for Rainbow Trout: Limitations in their use for prediction of environmental residues, *Environ. Sci. Technol.,* 19, 842, 1985.

38. **Quayle, O. R.,** The parachors of organic compounds, *Chem. Rev.,* 53, 439, 1953.

39. **Reid, R. C., Prausnitz, J. M., and Sherwood, T. K.,** *The Properties of Liquids and Gases,* 3rd Ed., McGraw-Hill, New York, 1977, Chap. 3.

40. **Hawker, D. W. and Connell, D. W.,** Octanol-water partition coefficients of polychlorinated biphenyl congeners, *Environ. Sci. Technol.,* 22, 382, 1988.

41. **Södergren, A.,** Solvent-filled dialysis membranes simulate uptake of pollutants by aquatic organisms, *Environ. Sci. Technol.,* 21, 855, 1987.

42. **Mackay, D.,** Correlation of bioconcentration factors, *Environ. Sci. Technol.,* 16, 274, 1982.

43. **Dobbs, A. J. and Williams, N.,** Fat solubility-A property of environmental relevance, *Chemosphere,* 12, 97, 1983.

44. **Opperhuizen, A., Van de Velde, E. W., Gobas, F. A. P. C., Liem, D. A. K., and Van der Steen, J. M. D.,** Relationship between bioconcentration in fish and steric factors of hydrophobic chemicals, *Chemosphere,* 14, 1871, 1985.

45. **Muir, D. G. G., Marshall, W. K., and Webster, G. R. B.,** Bioconcentration of PCDD's by fish: Effects of molecular structure and water chemistry, *Chemosphere,* 14, 829, 1985.

46. **Bruggeman, W. A., Opperhuizen, A., Wijbenga, A., and Hutzinger, O.,** Bioaccumulation of super lipophilic chemicals in fish, *Toxicol. Environ. Chem.,* 7, 173, 1984.

47. **Ho, N. F. H., Park, J. Y., Morozowich, W., and Higuchi, W. I.,** Physical model approach to the design of drugs with improved intestinal absorption, in *Design of Biopharmaceutical Properties Through Prodrugs and Analogs,* Roche, E. B., Ed., American Pharmaceutical Association Academy of Pharmaceutical Sciences, Washington, D. C., 1977, Chap. 8.

48. **Opperhuizen, A., Serne, P. and Van der Steen, J. M. D.,** Thermodynamics of fish/water and octan-1-ol/water partitioning of some chlorinated benzenes, *Environ. Sci. Technol.,* 22, 286, 1988.

49. **Harrison, R. and Lunt, G. G.,** *Biological Membranes,* 2nd Ed., Blackie, Glasgow, 1980, Chap. 5.

50. **Stein, W. D.,** Permeability for lipophilic molecules, in *Membrane Transport,* Bonting, S. L. and de Pont, J. H. M., Eds., Elsevier, Amsterdam, 1981, Chap. 1.

51. **Diamond, J. M. and Wright, E. M.,** Biological membranes: The physical basis of ion and non-electrolyte selectivity, *Ann. Rev. Physiol.,* 31, 581, 1969.

52. **Lieb, W. R. and Stein, W. D.,** Biological membranes behave as non-porous polymeric sheets with respect to the diffusion of non-electrolytes, *Nature,* 224, 240, 1969.

53. **Anderson, B. D.,** Thermodynamic considerations in physical property improvement through prodrugs, in *Physical Chemical Properties of Drugs,* Vol. 10, Yalkowsky, S. H., Sinkula, A. A., and Valvani, S. C., Eds., Marcel Dekker, New York, 1980, Chap. 7.

54. **Tulp, M. T. M. and Hutzinger, O.,** Some thoughts on aqueous solubilities and partition coefficients of PCB, and the mathematical correlation between bioaccumulation and physicochemical properties, *Chemosphere,* 10, 849, 1978.

55. **McKim, J., Schneider, P. and Vieth, G.,** Absorption dynamics of organic chemical transport across trout gills as related to octanol-water partition coefficient, *Toxicol. Appl. Pharmacol.,* 77, 1, 1985.

56. **Van de Waterbeemd, H.,** The theoretical basis for relationships between drug transport and partition coefficients, in *Quantitative Approaches to Drug Design,* Dearden, J. C., Ed., Elsevier, Amsterdam, 1983, 183.

57. **Mackay, D. and Hughes, A. I.,** Three parameter equation describing the uptake of organic compounds by fish, *Environ. Sci. Technol.,* 18, 439, 1984.

58. **Bruggeman, W. A., Martron, L. B. J. M., Kooiman, D. and Hutzinger, O.,** Accumulation and elimination kinetics of di-, tri- and tetra chlorobiphenyls by goldfish after dietary and aqueous exposure, *Chemosphere,* 10, 811, 1981.

59. **Spacie, A. and Hamelink, J. L.,** Alternative models for describing the bioconcentration of organics in fish, *Environ. Toxicol. Chem.,* 1, 309, 1982.

60. **Herbes, S. E. and Risi, G. F.,** Metabolic alternative and excretion of anthracene by *Daphnia pulex, Bull. Environ. Contam. Toxicol.,* 19, 147, 1978.

Chapter 6

BIOCONCENTRATION OF LIPOPHILIC AND HYDROPHOBIC COMPOUNDS BY AQUATIC ORGANISMS

Des W. Connell

TABLE OF CONTENTS

I. MECHANISM OF BIOCONCENTRATION

A. UPTAKE AND CLEARANCE ROUTES

Hansch and Fugita[1] in 1964 proposed that the mechanism for entry of chemicals into a cell from an external dilute solution was by diffusion. They noted previous work with *Nitella* cells which found that the rate of movement of an organic compound through cellular material was approximately proportional to the log K_{ow} of the compound. Later work by a variety of researchers has substantiated this suggestion. For example, Baughman and Paris[2] reviewed the available information on bioconcentration with microbes and concluded that the mechanism was similar to simple cell-to-water partitioning. Sondergren[3] found that with *Chlorella* the uptake mechanism was physical absorption, and the rate was the same as the rate of diffusion in water. Similarly, Kerr and Vass[4] concluded that unicellular organisms adsorb lipophilic compounds on the outer cell surface, and then the compound moves within the cell by simple diffusion. This uptake route is basically the entry route for oxygen from the dissolved oxygen in the water mass surrounding the organism.

Organisms at a higher level of organization exhibit related mechanisms of uptake. Crosby and Tucker[5] in 1971 found that the zooplankton species *Daphnia magna* took up DDT through the carapace. The results of Harding and Vass[6] on the bioconcentration of DDT from water by a marine zooplankton species could be explained by a single compartment model incorporating uptake directly from, and clearance to, water. Later work[7] on a variety of marine planktonic crustacea revealed that a similar single compartment model was adequate in these situations also. Supporting evidence for direct uptake from water through the carapace was obtained by Johnson et al.[8] in aquarium experiments. Thus, the oxygen uptake route seems to be the path followed by lipophilic chemicals with these organisms also.

A variety of other invertebrate organisms has been subject to bioconcentration experiments. For example, Wilkes and Weiss[9] found that for low DDT concentrations in the ambient water, physical diffusion was the uptake mechanism for the aquatic dragonfly nymph *Tetragoneura*, but at higher concentrations biological mechanisms may operate as well. With bivalves, Pasteels[10] has suggested that a micellar layer on the gill surface is responsible for the initial adsorption of hydrophobic compounds. In addition, the overall bioconcentration can be explained in terms of organism-to-water partitioning, according to the results of Geyer et al.[11] Ernst[12] has obtained results which indicate that the partitioning is between the lipid tissues in the organism, rather than the bulk of all of the tissues present in the organism, and water.

Infauna, such as oligochaete worms, present a different situation in comparison to the biota previously discussed, since the presence of sediment produces a more complex system surrounding the organism. The mechanism involved is not a simple partition, as illustrated by the results of Oliver.[13] However, Shaw and Connell[14] have suggested that uptake and clearance by oligochaetes occur via the sedimentary interstitial water in a somewhat similar manner to the uptake of lipophilic compounds from ambient water by other organisms.

Several researchers[15,16] demonstrated during the 1960s that lipophilic compounds could be both taken up and lost through the gills in fish. Experiments by Chadwick and Brockson[17] indicated the importance of an uptake route to fish directly from the ambient water. Then, as a result of extensive experiments in pools, Hamelink et al.[18] in 1971 proposed that the concentrations of lipophilic compounds in aquatic organisms, including fish, were generally controlled by water-to-organism partitioning. For fish, a two stage process, first, water to blood, and then blood to fats, was proposed. Reinert[19] later confirmed the importance of direct water uptake by fish which were also exposed to lipophilic compounds in food. Since that time, numerous results reviewed by Connell[20] have shown that bioconcentration is an organism-to-water partitioning process controlled by diffusion and related processes.

With fish and other aquatic organisms, the water-to-gill transfer is of critical importance in uptake. Holden[21] and other researchers have identified the gills, and not other body surfaces, as the site of uptake and clearance of unmetabolized chemicals. Murphy and Murphy[22] confirmed the gills as the prime site of uptake of lipophilic chemicals and evaluated the effects of water flow over the gills and environmental factors on this process. Importantly, they demonstrated a relationship between the uptake of DDT and the uptake of oxygen from water through the gills. With sodium linear alkylbenzene sulfonate and sodium alkylsulfate, which have different physicochemical properties from the range of lipophilic compounds considered above, Kikuchi et al.[23] found that the uptake route was also primarily through the gills with fish.

It has been previously mentioned that clearance of unmodified lipophilic chemicals in an organism occurs by release to the ambient water through the gills. If a compound is metabolized, it is usually converted into more polar substances by particular organs, and excreted in aqueous solution by some other route. Chemicals can enter organisms by bio-concentration or biomagnification, essentially through water or through food, respectively. However, once within an organism, chemicals taken up by either route are controlled by the same excretion and metabolic processes. Thus, for compounds which are essentially undegraded during the bioconcentration process, the chemical concentration within an organism is related to the clearance rate through the gills, essentially the process reverse to bioconcentration.

A set of common generalized routes for the uptake and clearance of lipophilic chemicals by aquatic organisms is apparent from the previous discussion and outlined in Figure 1. All aquatic biota, which obtain their supply of oxygen from that dissolved in the water mass, have special organs, e.g., gills, or can absorb oxygen by diffusion through their outer membranes. These surfaces are permeable to oxygen molecules, which makes them permeable to lipophilic chemicals which may bioconcentrate. Chemicals dissolved in the ambient water can move to the oxygen uptake membrane surface by diffusion, or by movement of the water mass, as induced by many aquatic organisms, e.g., operculum movement with fish. Passive diffusion through the oxygen uptake surface into the body tissues occurs with small organisms, such as micro-organisms, but with larger organisms the lipophilic chemicals diffuse into the circulatory fluid. Depending on its stability, the chemical is then accumulated in lipids in the organism, or metabolized and excreted. Thus, these processes together can be described as bioconcentration, and except for metabolism, are controlled by equilibrium exchange. Alternatively, consumption of a food organism leads to transfer of some of its xenobiotic chemical burden through the walls of the gastrointestinal tract of the consumer. The chemical is taken up by the circulatory fluid, leading to processes similar to those which occur with bioconcentration. Overall, these processes are described as biomagnification. However, the clearance of unmetabolized chemical originating from food then follows the route reverse to that described above for bioconcentration.

B. DEPOSITION AND BIODEGRADATION OF LIPOPHILIC CHEMICALS

In 1972 Hamelink et al.[18] proposed that the bioaccumulation of chlorinated hydrocarbons in aquatic systems was controlled by exchange equilibria between the water and organism. By analogy with partition chromatography and considerations of the lipophilic nature of the compounds, they also suggested that these lipophilic compounds were deposited in the body lipids of the biota. It is curious to note that earlier Grzenda et al.[24] reported that with DDT bioaccumulation in fish there was no apparent relationship between lipid and pesticide concentration in the tissues. But at the same time, mesenteric adipose tissues had among the highest lipid and pesticide concentrations observed.

A variety of different types of aquatic organisms has been shown to accumulate lipophilic compounds in body lipids and organs where metabolism occurs. For example, Connell[25] found that hydrocarbons accumulated in fish muscle tissues from a contaminated natural ecosystem according to the following equation

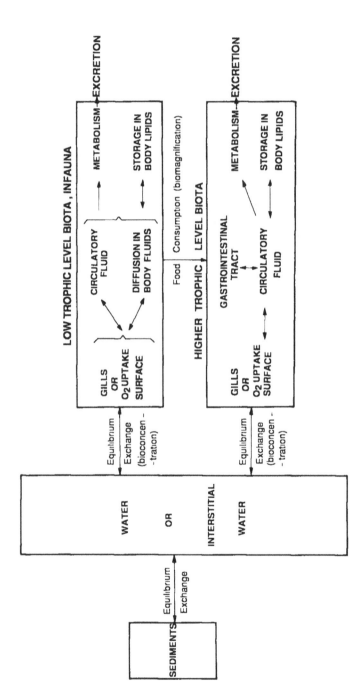

FIGURE 1. Diagrammatic representation of routes of uptake and clearance of lipophilic chemicals by aquatic biota.

$$[\text{percentage hydrocarbon}] = 1.086 \, [\text{percentage lipid}] - 1.23$$

This indicates a close relationship between lipid and hydrocarbon content of muscle tissue in the contaminated fish. Laboratory experiments by Muir et al.[26] have demonstrated that depot fat in fish contains the highest concentrations of dioxins apart from bile, which contained a high proportion of these compounds in a conjugated form.

With other types of aquatic organisms, there has been similar evidence indicating that lipophilic compounds accumulate in body lipids and in some organs. Fong[27] has demonstrated in laboratory experiments with clams that crude oil compounds were incorporated into the lipid components of body tissues. Solbakken et al.[28] have observed that the highest concentrations of radioactivity from bioaccumulated radioactive phenanthrene was in the hepatopancreas of clams. The freshwater filter feeder *Sphaerium corneum* has been found by Boryslawskyj et al.[29] to accumulate dieldrin from water into lipoidal tissues. Ernst[12] found that the amount of bioconcentration with mussels was proportional to the lipid content of the organs, supporting lipid as the final site of deposition. Also, Goerke[30] has reported that in laboratory experiments with the sand worm, various parts accumulated PCBs according to their lipid content. Connell[25] has pointed out that while a close relationship between hydrocarbon content and lipid content was observed for tissues within fish, the total lipid content was not related to the total hydrocarbon content for the set of individual fish he investigated. It was suggested that this was due to individual fish having different exposure periods to hydrocarbons in water. If aquatic organisms are exposed to an equivalent degree to the same lipophilic compound, then the observed K_B value calculated on a wet weight basis would be expected to be related to the organism's lipid content. Geyer et al.[31] have reported such a relationship with some species of fish and mussels. In addition, Geyer et al.[31] have suggested that those species which have high lipid contents will exhibit correspondingly higher K_B values.

Lipophilic compounds are stored in body lipids where little degradation occurs, but can be remobilized into the circulatory fluid and passed to organs where metabolism and degradation can occur (see Figure 1). On reviewing the available data, Zitko[32] comments that metabolic degradation is generally slower in aquatic, as compared to terrestrial organisms, and that the parent compound is often the main excretionary product with aquatic organisms.

The metabolic degradation of xenobiotic compounds can be divided into two phases: first, formation of metabolites, usually by action of mixed function oxidize (MFO) enzymes systems, and then conjugation with a variety of compounds, as summarized in Figure 2. Generally, there is the production of less toxic, more hydrophilic compounds, which can be excreted by the organism. MFO activity in organisms can be induced by exposure to lipophilic xenobiotic chemicals. Most xenobiotic compounds of environmental importance are powerful inducers of MFO activity. In some cases, the products of this detoxification process are harmful, e.g., with some polycyclic aromatic hydrocarbons (PAH) the oxides produce primary carcinogens.

II. QUANTITATIVE STRUCTURE-ACTIVITY RELATIONSHIPS IN BIOCONCENTRATION

A. GENERAL NATURE OF RELATIONSHIPS
1. Lipophilic Compounds

In the following two sections a distinction is made between lipophilic and hydrophobic compounds, and these are discussed separately. The following discussion applies generally to lipophilic compounds which do not degrade significantly in the time taken for bioconcentration to occur. These compounds have lipophilicity, measured by the log K_{ow} value, of from about 2 to about 7. Below about 2, compounds do not exhibit a marked affinity for

BIOLOGICAL INTERACTIONS WITH POLLUTANTS

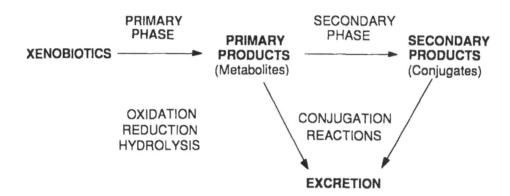

FIGURE 2.   Generalized biotransformation pathways for environmental xenobiotic compounds. (From Connell, D. W. and Miller, J. G., *Chemistry and Ecotoxicology of Pollution*, John Wiley & Sons, New York, 1984, 39. Copyright John Wiley & Sons. With permission.)

lipid, while above 7 Connell and Hawker[33] have suggested that lipid solubility may fall and other effects may occur leading to deviations from the behavior observed with the lipophilic group. The general discussion outlined below does not apply to compounds outside of the lipophilic group as defined above. The group of hydrophobic compounds includes the lipophilic compounds with log K_{ow} values between 2 and 7, as well as those compounds with log K_{ow} greater than 7, and this overall group is discussed in the next section.

Hamelink et al.[18] suggested in 1971 that bioconcentration of lipophilic chemicals in lentic environments were controlled by exchange equilibria between the ambient water and body fats. This mechanism implied that the octanol-to-water partition coefficient, and the related water solubility value, would be useful characteristics of a compound to explain its bioconcentration in aquatic organisms. Neely et al.,[34] Kennaga and Goring,[35] Lu and Metcalfe,[36] Ernst,[37] and others in later years found that log K_B was linearly related to these characteristics with a variety of aquatic organisms. Connell[20] has listed relationships established between log K_B and log K_{ow} for fish, molluscs, invertebrates, and microorganisms, as well as relationships between log K_B and log S for fish and molluscs. These relationships have the general form

$$\log K_B = a \log K_{ow} + b \quad \text{or} \quad K_B = 10^b K_{ow}^a \tag{1}$$

where a and b are empirically developed constants.

Mackay[38] has developed the following treatment for the theoretical nature of the relationship between K_B and K_{ow} for compounds which do not degrade significantly during the time of bioconcentration, and for the situation when equilibrium is established between the organism and water. Fugacity is a valuable property for investigating substances at equilibrium between different phases, since all fugacities are equal at this stage. If a fish or any other aquatic organism where bioconcentration occurs is considered to consist of a number of phases (1,2,3, etc.), then at equilibrium

$$f_w = f_1 = f_2 = f_3 = f_i$$

where f_w is fugacity of the solute in the surrounding water, f_1, f_2, f_3 etc., is the fugacity in the phases within the organism, and f_i is the fugacity in any phase.

The value of f_i can be calculated using a Raoult's Law Convention from

$$f_i = x_i \, Y_i f_R = c_i \, v_i \, Y_i f_R \qquad (2)$$

where x_i is the mole fraction of solute which is small so that it has a negligible effect on the phase molar volume v_i, c_i is the compound concentration, Y_i is the activity coefficient, f_R the reference fugacity. Now first consider the organism. The amount of solute (m_i moles) in each phase (1,2,3, etc.) comprising the organism is

$$m_i = c_i \, y_i \, v$$

where y_i is the volume fraction of the phase in the organism, and v is the total volume of the organism. Thus the average biotic concentration (C_B) is

$$C_B = \Sigma \, m_i/v = \Sigma \, c_i \, y_i$$

substituting for c_i from Equation 2, then

$$C_B = (f_i/f_R) \, \Sigma \, (y_i/Y_i \, v_i) \qquad (3)$$

Turning to considerations of the surrounding water, we have by analogy with Equation 2

$$f_W = C_W \, v_W \, Y_W \, f_R$$

where C_W is solute concentration in water, v_W the molar volume of water, and Y_W the activity coefficient of the solute in water. Thus,

$$C_W = f_W/(v_W \, Y_W \, f_R)$$

Using this equation and Equation (3) and noting that $f_i = f_W$ then

$$K_B = C_B/C_W = v_W \, Y_W \, \Sigma \, [y_i/(Y_i \, v_i)] \qquad (4)$$

The K_{ow} values were considered by Mackay[38] as follows:

$$K_{ow} = C_o/C_W = (x_o/v_o)/(x_W/v_W)$$

where C_o is the concentration of solute in octanol, C_W the concentration in water, x_o and x_W the mole fractions of solute in octanol and water, respectively, and v_o and v_W the molar volumes of octanol and water, respectively. But at equilibrium the fugacities are equal, so by analogy with Equation (2)

$$f = x_o \, Y_o \, f_R = x_W \, Y_W \, f_R$$

thus $\qquad K_{ow} = Y_W \, v_W/Y_o \, v_o \qquad (5)$

The relationship of K_B to K_{ow} can be evaluated considering the ratio K_B/K_{ow} derived from Equations 4 and 5. Thus

$$K_B/K_{ow} = \Sigma[(y_i/(Y_i \, v_i)] \, Y_o \, v_o$$

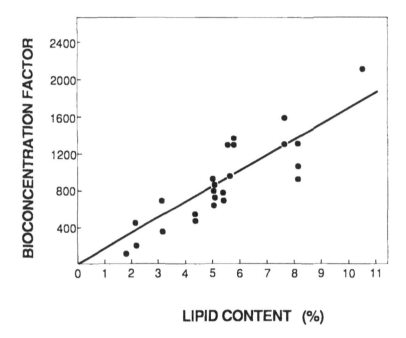

FIGURE 3.   Relationship of lipid content K_B for trichlorobenzene bioconcentrated in eight species of fish. (From Geyer, H., Scheunert, I., and Korte, F., *Chemosphere*, 14, 549, 1985. Copyright Pergamon Press. With permission.)

Mackay[38] has suggested that the lipid phase in aquatic organisms is the dominant concentrating phase. The aqueous phases have little capacity to take up these highly lipophilic compounds. This is reflected by the results reported in the previous section indicating that lipophilic compounds are deposited in fats within organisms. Thus it could be expected that

$$K_B/K_{OW} = y_L(Y_o\ v_o)/(Y_L\ v_L) \tag{6}$$

where y_L is the fraction of lipid in the organism. Although lipid composition may vary with different organisms, v_L would be expected to be effectively constant, as would v_o. This would lead to an effective constant ratio for v_o/v_L which would apply particularly for organisms with similar lipid composition characteristics; and also, where the lipid and octanol have similar properties, which would be expected in most cases. The ratio of activity coefficients (Y_o/Y_L) is effectively constant. This means that for organisms of different lipid contents (y_L variation), for the same compound (K_{OW} is constant) the K_B value will vary with the lipid content of the organism. Such a relationship has been observed by Geyer et al.[31] with different species of fish, as shown in Figure 3. For this data on trichlorobenzene the equation to the regression line (significant at the 99.9% level) was

$$K_B = 166\ L$$

where L is the percentage of lipid in the organism. On the other hand, for organisms of the same lipid content (y_L is constant), K_B/K_{OW} should be constant for lipophilic compounds, and K_B itself should be constant for a single compound. Comparison of triolein-to-water partition coefficients with octanol-to-water partition coefficients has been made by Chiou[39] based on the assumption that triolein is a good representative of organism fat. He concluded that K_{OW} values were good predictors of the triolein-to-water partition coefficient and the bioconcentration factor. This suggests that octanol is an acceptable surrogate for organism fat. From Equation 6, if octanol is a perfect surrogate for organism lipid then

$$Y_o \, v_o / Y_L \, v_L = 1$$

thus

$$K_B = K_{OW} \, y_L \quad \text{and} \quad \log K_B = \log K_{OW} + \log y_L \tag{7}$$

This equation can be compared with the form of the observed empirical equations (Equation 1), i.e.,

$$\log K_B = a \log K_{OW} + b$$

It can be seen that under these circumstances

$$b = \log y_L$$

The constant a should be unity in all cases if octanol is a perfect surrogate for organism lipid. The deviation from unity reflects how much octanol does, in fact, differ from organism lipids.

Some different types of graphs which should be theoretically obtained under these circumstances are shown in Figure 4. In fact, these relationships have been shown to extend over the $\log K_{OW}$ range from about 2 to about 7. In the situation where values of K_B are expressed in terms of the organism lipid weight, all points for all organisms would be expected to lie on a line with slope of unity and $\log K_B$ intercept of zero (see Figure 4). The graphs to be expected where K_B is expressed in wet weight terms and the organisms have different fat contents are also illustrated in Figure 4. It is noteworthy that the intercept on the $\log K_B$ axis should always be negative. Of course, there are many different factors which may lead to deviations from these perfect relationships.

The other physicochemical parameter often used in quantitative structure-activity relationships (QSAR) for bioconcentration is water solubility (S). The nature of the K_B to S relationship can be better understood by development of the $\log K_B$ to $\log K_{OW}$ relationship outlined above. The investigations of a number of authors have established direct relationships between K_{OW} and S, which were discussed in Chapter 2. It is noteworthy that it is necessary to correct the solubility of solids, since solids have lower solubility than their super cooled liquids. Mackay,[38] after making this correction, found the following relationship

$$\log K_{OW} = 3.25 - \log C_l$$

where C_l is the water solubility (mole m^{-3}) of liquids or corrected solids. By substituting this expression into Equation 1 the $\log K_B$ to $\log C_l$ relationship is obtained

$$\log K_B = (3.25 \, a + b) - a \log C_l$$

This means that $\log K_B$ should increase as $\log C_l$ declines. However, the numerical value of the slope of this relationship, when octanol perfectly represents organism lipid, is unity. The intercept on the $\log K_B$ axis, in this case, is a more complex combination of factors. If constant a is unity, it should be $+3.25$ when K_B is expressed in terms of lipid weight. If constant a is unity and K_B is expressed in terms of wet weight with an organism having a fat content of 1% (y_L 0.01), then the intercept should be at 1.25. The intercept on the $\log K_B$ axis can be similarly estimated for other organism lipid contents when K_B is expressed in wet weight terms.

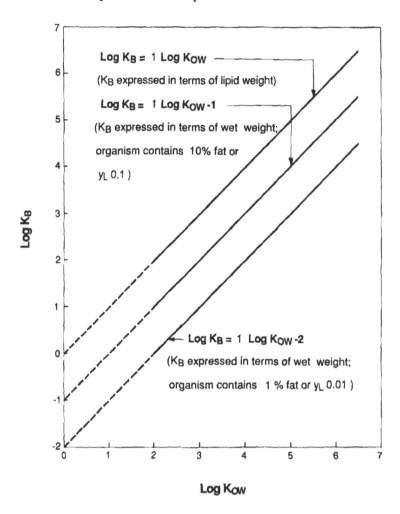

FIGURE 4.   Theoretical graphs of the log K_B to log K_{ow} relationship where octanol perfectly represents the organism lipid.

It should be kept in mind that these relationships are only applicable at equilibrium. In many cases equilibrium is not attained due to insufficient exposure time. The kinetics of bioconcentration and the times to equilibrium are discussed in a later section. Another factor which may play a role in the relationships previously discussed is the stereochemistry of the molecules involved. This is discussed in Section III of this chapter.

2. Hydrophobic Compounds
a. The Relationships Between Bioconcentration and the Octanol-to-Water Partition Coefficient and Aqueous Solubility

In the previous section lipophilic compounds were defined as those compounds with log K_{ow} values between 2 and about 7. Compounds with log K_{ow} values above about 7 are often referred to as "superlipophilic." This suggests that superlipophilic compounds have lipid solubility properties in excess of those with log K_{ow} values from 2 to about 7. However, Dobbs and Williams[40] point out that compounds with log K_{ow} greater than about 6 in fact exhibit declining fat, or lipid, solubility. Similarly, Connell and Hawker[33] suggest that the decline in fat solubility in this range has an important impact on the log K_B-to-log K_{ow} relationship. Thus, "superlipophilic" is an inappropriate term to use for this group of compounds. These compounds have extreme hydrophobic properties, as indicated by their water solubility, which is lower than that of the lipophilic compounds with log K_{OW} values between 2 and about 7. As a result, it is suggested that "superhydrophobic" is a more appropriate term than "superlipophilic" for this group of compounds.

When the lipophilic compounds and superhydrophobic compounds are treated as a single group quite different overall relationships are obtained than those which have been described for the lipophilic compounds alone. Thus the lipophilic compounds, which also have hydrophobic properties, and the superhydrophobic groups together are referred to here as hydrophobic compounds. This means that hydrophobic compounds have log K_{OW} values which range from about 2 to the highest values of log K_{OW} which can be assigned. Hansch[41] in 1969 suggested that the relationship between the hydrophobic character of molecules and their induced biological effects was nonlinear. He suggested that overall a parabolic relationship was to be expected having the form

$$\log R = k_A (\log K_{OW})^2 + k_B(\log K_{OW}) + k_C$$

where R is an expression for a biological response, and k_A, k_B, and k_C are empirical constants.

It was pointed out that for moderately hydrophobic compounds this relationship would approach a linear form, but very hydrophobic compounds would give a reduced response resulting in a parabolic relationship overall. The reason for a fall in biological response was believed to be due to the highly lipophilic compounds lacking the ability to penetrate and diffuse from lipophilic barriers. Bioaccumulation can be seen as the initial biological effect of a compound on an organism, and so the factors mentioned above are relevant to this discussion.

The general principles involved in bioconcentration of lipophilic compounds have been fairly well established, as described in the previous section. Also, there is a substantial amount of information available on the bioconcentration of these compounds by fish and some other organisms, as will be described later. But limited information is available on the full range of hydrophobic compounds, particularly those in the superhydrophobic range. Data obtained by Sugiura et al.[42] in 1978 indicated a fall in bioconcentration with compounds having log K_{OW} values greater than about 6. Similar results have been obtained by Konemann et al.,[43] Muir et al.,[44] Opperhuizen et al.,[45] Bruggeman et al.,[46] and Anliker et al.[47] In these reports the bioconcentration of compounds having large molecules and log K_{OW} values greater than 6 had bioconcentration substantially less than would be expected from the linear equation, Equation 1. This equation is based on lipophilic compounds with log K_{OW} values from 2 to about 7. In addition, Brooke et al.[48] noticed a sharp decline in bioaccumulation of chlorinated hydrocarbons at molecular weights in excess of about 300.

Connell and Hawker[33] have reviewed the information available. They found that for the bioconcentration of chlorinated hydrocarbons and related compounds by fish, at equilibrium the log K_B values were related to the log K_{OW} values by a polynomial equation as follows:

$$\log K_B = 6.9 \times 10^{-3}(\log K_{OW})^4 - 1.85 \times 10^{-1}(\log K_{OW})^3$$
$$+ 1.55(\log K_{OW})^2 - 4.18 \log K_{OW} + 4.79 \tag{8}$$

The curve described by this equation resembles a parabola over the hydrophobic chemical log K_{OW} value range. It has a maximum log K_B value of 4.61 which occurs with compounds having a log K_{OW} value of 6.7 (see Figure 5).

For the data used to obtain this expression with lipophilic compounds, considered to have log K_{OW} values between 3 and 6, the equation to the linear regression line is

$$\log K_B = 0.94 \log K_{OW} - 1.00 \qquad \text{ue17}$$

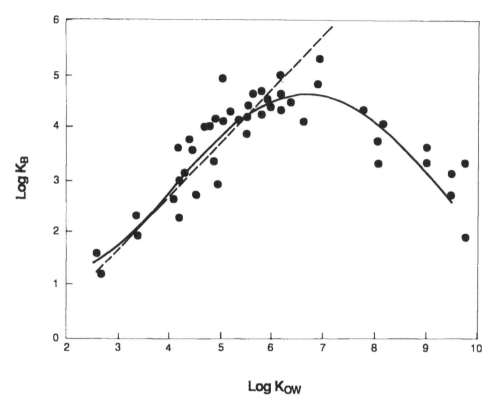

FIGURE 5. The relationship between log K_B and log K_{OW} for bioconcentration of hydrophobic compounds (log K_{OW} about 3 to about 10; represented by the full line) and lipophilic compounds (the broken line) and the original data represented by the full circles. (Adapted from Connell, D. W. and Hawker, D. W., *Ecotoxicol. Environ. Saf.*, in press. Academic Press, New York.)

According to the discussion outlined previously for lipophilic compounds, if octanol perfectly represents organism fat then

$$\log K_B = \log K_{OW} + \log y_L$$

If fish have an average lipid content of 10%, then

$$\log K_B = \log K_{OW} - 1.00$$

Thus, there is reasonable agreement over the approximately linear section of the polynomial curve between the actual equation obtained and the equation theoretically expected for lipophilic compounds (see Figure 5). This means that for lipophilic compounds, treated as a separate set of compounds with log K_{OW} values from 3 to 6, and for the full set of hydrophobic compounds, taken over a wider range of log K_{OW} values from about 3 to 9.5, there is reasonable agreement between the equations obtained (see Figure 5).

The linear regression equation developed for lipophilic compounds described in the previous section was developed on the basis that octanol was a perfect surrogate for organism lipid, or acceptably close to it. It has been suggested by Connell and Hawker[33] that octanol is not a good surrogate for superhydrophobic compounds.

For lipophilic compounds where octanol is a perfect surrogate for organism lipid or close to this situation

$$C_B/C_W = K_B = K_{OW} = C_O/C_W$$

where C_B, C_O and C_W are maximum biotic, octanol, and aqueous solubilities, respectively.

For superhydrophobic compounds at equilibrium, since K_B is less than K_{OW}, then it follows that C_B is less than C_O. The most apparent difference between octanol and lipid is molecular size. The solubility of hydrophobic solutes, taking into account solvent molecular size, may be treated by an adaption of the Flory-Huggins Theory, as described by Chiou,[39] and approximated by equations of the general form

$$\log C_X = -\log v_{SOL} - (1 - v_{SOL}/v_X^*)/2.303 - X/2.303 \qquad (9)$$

where C_X is the solubility in either octanol or lipid, v_{SOL} the molar volume of the solute, v_X^* the molar volume of water saturated solvent, X the Flory-Huggins interaction parameter, which is a measure of solute-solvent compatibility. The molar volume for water-saturated octanol[49] is approximately 127 cm^3 mol^{-1}, while that for lipid (as represented by glyceryl triolate) is about 7 times larger.[39] It could be expected that solute molar volume and molecular size would roughly increase with hydrophobicity for related chemicals. The second term of Equation 9, i.e., $(1 - v_{SOL}/v_X^*)/2.303$ is larger for lipid than for octanol as a solvent; therefore, this tends to make lipid an increasingly poor solvent relative to octanol with increasing solute size. For solute molecules whose molar volume is roughly that of octanol, within a factor of 2 or 3, octanol and lipid solubilities are roughly equal because this tendency is apparently balanced by increasing solute incompatibility, $X_O/2.303$, in octanol. The incompatibility term is negligible for lipid since v_{SOL} is much greater in size than v_L. Only for relatively large solutes, superhydrophobic compounds, will the solute incompatibility term, $X_L/2.303$ from Equation 9 have an effect for lipid. With such compounds and increasing $X_L/2.303$, the term means that lipid concentration (C_L), and hence C_B, may become less than C_O.

If this is the case then the bioconcentration factor on a lipid basis for superhydrophobic compounds may be expressed as

$$K_B = C_L/C_W < C_O/C_W = K_{OW}$$

The polynomial expression shown in Equation 8 can be extended to derive the log K_B to aqueous solubility relationship. Connell and Hawker[33] report the following relationship between log K_B and log S_W, where S_W is aqueous solubility in moles m^{-3}

$$\log S_W = -1.24 \log K_{OW} + 4.01 \qquad (10)$$

When this equation is used to substitute for log K_{OW} in Equation 8 the following relationship is obtained:

$$\log K_B = 2.91 \times 10^{-3}(\log S_W)^4 + 5.02 \times 10^{-2}(\log S_W)^3$$
$$+ 1.23 \times 10^{-1}(\log S_W)^2 - 4.15(\log S_W) + 1.98$$

These relationships apply at equilibrium, and the influence of kinetic factors as well as stereochemical factors are discussed in following sections in this chapter. But in fact, limited data are available on superhydrophobic compounds to allow these relationships to be unequivocally established. Most relationships have been established for the lipophilic group of compounds.

The equation above and Equation 8 indicate that the superhydrophobic compounds have the same log K_B values as the corresponding lipophilic compounds with log K_{OW} values less than 6.7 (see Figure 5). However, an important difference between these two sets of chemicals should be noted. With superhydrophobic chemicals, the concentrations of chemical in both water and lipid are lower, in many cases orders of magnitude lower than those for the corresponding lipophilic chemicals. In practical terms this means that the K_B values of superhydrophobic chemicals will present particular difficulties in experimental determination. It is likely that the concentrations of chemicals with this group of substances are too low to be measured by techniques used with lipophilic compounds, and that special techniques would be needed to obtain a satisfactory result.

b. Maximum Biotic Concentration Attained by Organisms

Figure 5 illustrates the relationship between log K_B and log K_{OW} for hydrophobic chemicals which are bioconcentrated by fish. Similar results would be expected with other aquatic organisms. From this graph it can be seen that compounds with log K_{OW} values of 5.0 and 8.3 should have the same K_B value. However, the C_B and C_W values yielding these K_B values are greatly different, as described above.

The nature of the relationship between C_B and K_{OW} has been established by Connell and Hawker[33] as follows. Since

$$K_B = C_B/C_W$$

When C_B is at a maximum (C_{BM}) the C_W will be at the maximum aqueous solubility (S_W). Then

$$\log C_{BM} - \log S_W = \log K_B$$

and from Equation 8

$$\log C_{BM} = 6.9 \times 10^{-3}(\log K_{OW})^4 - 1.85 \times 10^{-1}(\log K_{OW})^3$$
$$+ 1.55(\log K_{OW})^2 - 4.18(\log K_{OW}) + 4.79 + \log S_W \qquad (11)$$

Combining Equations 10 and 11 we get

$$\log C_{BM} = 6.9 \times 10^{-3}(\log K_{OW}^4) - 1.85 \times 10^{-1}(\log K_{OW}^3)$$
$$+ 1.55(\log K_{OW}^2) - 5.42 \log K_{OW} + 8.80 \qquad (12)$$

The curve described by this equation is presented in Figure 6. This shows that log C_{BM} decreases slowly with increasing log K_{OW} until a log K_{OW} value of 5.5 is reached. For more hydrophobic molecules log C_{BM} decreases more rapidly, with C_{BM} for a compound with log K_{OW} of 9 reaching only 10^{-4} mole m^{-3}. The major repository for chlorinated hydrocarbons in fish is in the lipid tissues; therefore, the maximum biotic concentration is closely related to maximum lipid solubility. Thus, it would be expected that the relationship in Figure 6 would represent the general form of the relationship between log K_{OW} and lipid solubility.

3. Relationships with Infauna

The relationships previously described apply for biota which have free exchange with the ambient water mass. Benthic fauna, such as polychaetes, oligochaetes and other invertebrates which reside within aquatic sediments, do not have this free exchange. In these cases the situation regarding bioaccumulation is complicated by the presence of sediments. Bioaccumulation with infauna has been demonstrated with polychaetes and oligochaetes by Shaw and Connell,[14] Courtney and Langsdon,[50] Oliver[51] and Oliver.[52]

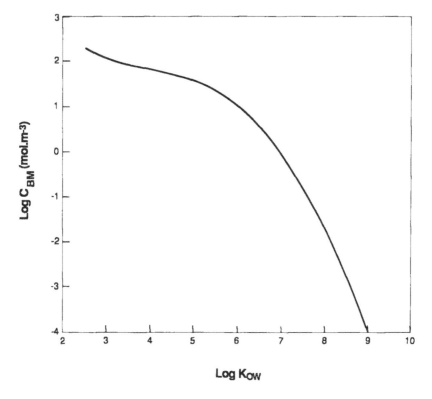

FIGURE 6. The relationship between log C_{BM} (maximum biotic concentration) and log K_{OW} derived from the empirically developed equation relating log K_B and log K_{OW} for fish. (Adapted from Connell, D. W. and Hawker, D. W., *Ecotoxicol. Environ. Saf.*, in press. Copyright Academic Press.)

FIGURE 7. Infauna can be considered to be part of a sediment, water and organism system, as illustrated above.

An approach to understanding some aspects of this system can be obtained by considering the infauna to be part of a sediment, water, and organism system, as illustrated in Figure 7. The essential difference between this situation and that of the organisms in the ambient water mass is that the availability of a chemical is closely controlled by the sediment-to-water equilibrium. It should be kept in mind that the rates of movement and availability of a chemical may play an important role in regulating the concentrations which occur in these three phases. For example, the mass of chemical available in the interstitial water is very small, since the mass of interstitial water is low, as is the concentration. These factors and others are not taken into account in the following treatment, and may be important.

First, the sediment-to-water system at equilibrium can be described by the Freundlich equation discussed in Chapter 4.

$$C_S = K_{SW} \, C_W^{1/n}$$

where C_S is sediment concentration, C_W water concentration, K_{SW} the sorption coefficient, and n a nonlinearity constant. In most environmental situations, the concentration of xenobiotic compounds is low and n is approximately unity.

However,

$$K_{SW} = K_{OC} \, f_{OC}$$

where K_{OC} is the sorption coefficient in terms of organic carbon, and f_{OC} is the fraction of organic carbon in the sediment. Also

$$K_{OC} = x \, K_{OW}^m$$

where x is the proportionality constant and m is the nonlinearity constant. Thus

$$K_{SW} = x \, K_{OW}^m \, f_{OC} \qquad (13)$$

and from the Freundlich equation, considering n to be effectively unity,

$$C_S = x \, K_{OW}^m \, f_{OC} \, C_W$$

Now, concerning the concentration of compound in the organism in relationship to the water: if the bioaccumulation occurs by transfer from water at equilibrium, it would be expected that the bioconcentration factor (K_B) would be

$$K_B = y_L \, K_{OW}^a = C_B/C_W$$

This is derived from Equation 7, where y_L is the lipid fraction in the organism and K_{OW} the partition coefficient of the compound which is being bioconcentrated, and constant a the nonlinearity constant. Thus

$$C_B = C_W \, y_L \, K_{OW}^a$$

Turning to the calculation of bioaccumulation factors, if the bioaccumulation factor (BF) is calculated on the basis of the sediment concentration then

$$BF = C_B/C_S = (C_W \, y_L \, K_{OW}{}^a)/(x \, K_{OW}{}^m \, f_{OC} \, C_W)$$

$$BF = C_B/C_S = (C_W \, y_L \, K_{OW}^a)/(x \, K_{OW}^m \, f_{OC} \, C_W)$$

and

$$BF = y_L/(x \, f_{OC}) \, K_{OW}^{a-m} \qquad (14)$$

The factor $y_L/x \, f_{OC}$ in this equation for the bioaccumulation factor is constant, and the constants a and m both usually approach unity; this suggests that BF should not vary with the K_{OW} value of the compounds being bioaccumulated, or should have a weak dependence on K_{OW}. It should principally reflect, and be equivalent to, the distribution of a compound between the similar phases of organic carbon and organism fat. These two phases are both lipoid in nature; thus, distribution between them would not be expected to be influenced by the octanol water partition coefficient. But it should be kept in mind that this applies at equilibrium, and other factors previously mentioned may have an influence on the process.

On the other hand, Shaw and Connell[14] have expressed the bioaccumulation factor for infauna in terms of the interstitial water concentration, thus regarding the bioaccumulation

TABLE 1
Characteristics of the Relationships Between Log K_B and Log K_{OW} for Bioconcentration of Lipophilic Compounds by Microorganisms

Constant a	Constant b	Number of values	r^2	Basis for K_B	Compound types	Organisms	Ref.
0.70	−0.26	8	0.93	Wet wt.	Pesticides	Alga	53
0.68	+0.16	41	0.81	Wet wt.	Diverse organic	Alga	54
1.08	−1.30	8	0.98	Dry wt.	Condensed ring aromatics	Mixed microbial	55
0.91	−0.36	14	0.95	Dry wt.	Diverse organic	Mixed microbial	2
0.46	+2.36	8	0.83	Wet wt.	Hydrocarbons and chlorohydrocarbons	Alga	56
0.36	+2.1	28	0.91	Wet wt.	Mainly chlorohydrocarbons	Alga	57

Note: Relationships between log K_B and log K_{OW} takes the form log K_B = a log K_{OW} + b.

process as bioconcentration from interstitial water. Considering the water concentration and using the Freundlich Equation in the form likely to apply to low concentrations in the environment, i.e., n is unity, then

$$C_S = K_{SW} C_W$$

and

$$K_B = C_B/C_W = C_B K_{SW}/C_S \qquad (15)$$

Values obtained using this expression with infauna should be comparable to those obtained in the normal way with fish. This relationship can be expressed in terms of K_{OW} from Equation 13. Thus

$$K_B = x C_B K_{OW}^m f_{OC}/C_S$$

B. QSARS WITH VARIOUS ORGANISM GROUPS
1. Microorganisms

A summary of observed relations between log K_B and log K_{OW} for lipophilic compounds is contained in Table 1. These data indicate a relationship between log K_{OW} and K_B with a wide range of microorganisms from algae to bacteria. However, there is a considerable degree of variability in the nature of the relationships found. It is noteworthy that many experiments have been conducted over short time periods, and these may not have had sufficient time to allow the establishment of equilibrium. Mallhot[57] has investigated algal bioaccumulation and uptake rates of various organic compounds. A variety of other factors may also contribute to errors in the determination of the constants in the log K_B-to-log K_{OW} relationship.

Possibly, the equation obtained by Baughman and Paris[2] is the most satisfactory, since this involved an extensive review of the literature and the use of a broad range of data from many experiments. These authors point out that the partition mechanism with microorganisms is a passive process since it occurs to similar extents with live and dead cells. The data obtained are consistent with a simple partition equilibrium between organism and water.

The Baughman and Paris[2] relationship has a constant a value of almost unity (0.91), consistent with octanol being a good surrogate for microorganism lipid with lipophilic compounds. In addition, constant b has a value of −0.36, which suggests a lipid fraction of about 44% of the dry weight. Thus, theoretical interpretation of the log K_B to log K_{OW} relationship is in support of the Baughman and Paris relationship.

TABLE 2
Characteristics of the Relationship Between Log K_B and Various Factors for Bioconcentration of Lipophilic Compounds by *Daphnia pulex*

Factor	Constant a	Constant b	Number of values	r^2	Basis for K_B	Compound types	Ref.
log K_{OW}	0.75	−0.44	7	0.85	Wet wt.	PAH	59
$^3X_c^v$	4.82	1.28	6	0.94	—	PAH	60
log K_{OW}	0.90	−1.32	22	0.96	Wet wt.	Diverse organic	61

Note: Relationship between log K_B and various factors takes the form log K_B = a (factor) + b.

The physicochemical characteristics most used after log K_{OW} in many QSARs is water solubility. Geyer et al.[58] found the following relationship for bioconcentration of a wide range of types of organic compounds by the alga, *Chlorella*.

$$\log K_B = -0.46 \log S + 4.549$$

$(r^2 = 0.76$ with 34 chemicals and S is in $\mu g\ L^{-1})$

On the other hand, Steen and Karickhoff[55] used 14 different microbial sources and investigated the bioconcentration of condensed ring aromatics and obtained the following relationship

$$\log K_B = -0.97 \log S_W - (0.022/2.303)(mp - 25) - 2.33$$

$(r^2 = 0.964$ with 8 chemicals and S_W is in mole fraction solubility)

2. Crustacea

A limited amount of data is available on QSARs for bioconcentration of lipophilic compounds by Crustacea. Some of this information is summarized in Table 2. The two relationships for log K_B to log K_{OW} are reasonably consistent. However, the relationship derived by Hawker and Connell[61] involves results from a range of experiments and has the advantage of utilizing more recent values for log K_{OW}. This suggests it may be the more accurate relationship for lipophilic compounds. In this relationship constant a approaches unity, as expected if octanol is a good surrogate for organism lipid. Also, the value of −1.32 for constant b indicates an organism fat content of 5%, which is a reasonable value.

A variety of other data is available on Crustacea, which is less amenable to interpretation as described above. For example, Zhang et al.[62] found that generally PCBs bioconcentration in *Daphnia magna* increased with decreasing water solubility. In investigations of lipophilic compounds derived from a discharge, Gossett et al.[63] found that concentrations in Crustacea and other organisms were positively correlated with log K_{OW}.

Some results are available on the accumulation of lipophilic xenobiotic compounds from sediment by Crustacea in the infauna. For example, Clark et al.[64] found that fiddler crabs accumulated PCB bound to sediments up to concentrations comparable with those existing in sediment. This is in accord with the previous discussion on bioaccumulation by infauna.

3. Poly- and Oligochaetes

Poly- and oligochaetes are infauna; the principles underlining QSAR relationships with this group are discussed in this chapter, Section II.A.3. Oliver[51,52] has carried out a set of bioaccumulation experiments in aquaria with oligochaete worms using chlorohydrocarbons. When the bioaccumulation factors (BF = C_B/C_S) were calculated the results shown in Figure 8 were obtained for the plot against the log K_{OW} values. Previously Equation 14, which

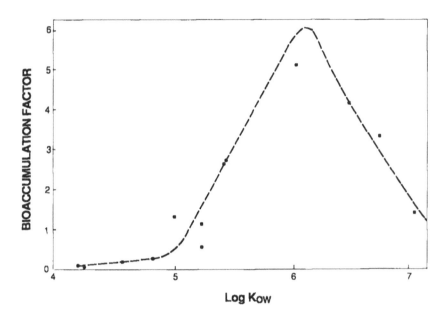

FIGURE 8. Plot of bioaccumulation factor (C_B/C_S, both in terms of dry weight) for the bioaccumulation of chlorohydrocarbons from sediments by oligochaetes. (From Oliver, B. G., *Can. J. Fish. Aquatic Sci.*, 41, 878, 1984. Copyright Government of Canada. With permission.)

predicted that BF would be independent of, or weakly dependent on, log K_{OW}, was developed for infauna. But with the K_B values of aquatic organisms a range of values is usually obtained, over the log K_{OW} range of 4 to 7, which spans several orders of magnitude and attains numerical values up to 10^5. However, here (see Figure 8) it is much less, and the relationship of BF to log K_{OW} is not linear.

When K_B is calculated according to water concentrations obtained using the Freundlich Equation, using the same set of data as reported by Oliver,[51] then Connell et al.[65] obtained the plots illustrated in Figure 9.

It is interesting to note that a polynomial equation fits the full set of data, which is in accord with previous discussion relating to hydrophobic compounds in this chapter, Section II.A.2. However, the most significant relationship for QSARs is for lipophilic compounds, which have log K_{OW} between 4.4 and 6.4. The linear regression line in this range has the following equation:

$$\log K_B = 2.13 \log K_{OW} - 7.53 \qquad (r = 0.97, \quad n = 22)$$

This is a significant relationship, but in previous sections on other organisms the most appropriate equation for bioconcentration of lipophilic compounds was similar to that derived from Equation 7. The comparison to this equation, constant a, 2.13, would be expected to be about 1.00, and the value -7.53 for constant b would represent a negligible lipid content. These deviations from the expected relationships outlined above may be due to kinetic factors which do not allow the system to reach equilibrium to be applicable. Possibly, colloids present in the interstitial water may complicate the situation, or other factors not presently apparent may be operating.

Later Oliver[52] produced another set of data based on a similar range of compounds and somewhat similar aquarium experiments. However, in these experiments the concentrations of the compounds in interstitial water was measured. If these observed C_W values are used to calculate K_B utilizing the observed concentrations in the worms, then plot C in Figure

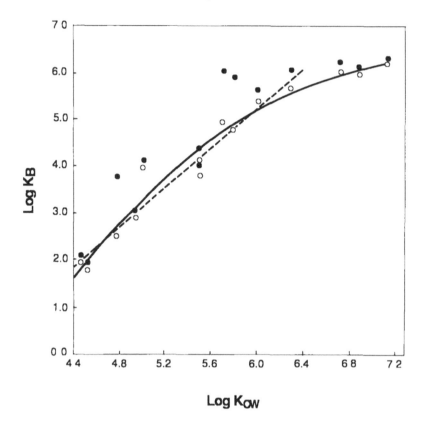

FIGURE 9. The relationship between log K_B and log K_{ow} when log K_B is derived from calculated water concentrations using data reported by Oliver.[51] The full line represents the polynomial equation of best fit and the broken line the linear regression line of best fit for log K_{ow} values between 4.4 and 6.4. (Adapted from Connell, D. W., Bowman, M., and Hawker, D. W., *Ecotoxicol. Environ. Saf.*, in press. Copyright Academic Press.)

10 is obtained. This suggests that log K_B is independent of log K_{OW}, but Connell et al.[83] took these observed C_W values and corrected them for the presence of colloids, thereby obtaining plot A. These authors also calculated new C_W values from the measured concentrations in the sediments (C_S) using Karickhoff's equation to give plot B. Both plots A and B have slopes close to unity and intercepts on the log K_B axis within a reasonable range of possible values in accord with lipid content.

These data support the bioaccumulation mechanism outlined in this chapter, Section II.A.3. and indicate that log K_{OW} may be a reasonable predictor of log K_B if this is based on interstitial water concentrations.

4. Fish

A substantial body of information is available on QSARs for bioconcentration of xenobiotic compounds by fish. Information on the log K_B to log K_{OW} relationship for lipophilic compounds is summarized in Table 3. An example of the plots of data used to obtain these relationships is shown in Figure 5. Many of the relationships described in Table 3 are inaccurate for one or more of the following reasons:

Inaccurate data — The values of log K_B and K_{OW} used in many studies have been later shown to be inaccurate. Thus, it could be expected that over time, values would be subject to checks and further examination; as a result, the accuracy of values generally would be expected to be better in more recent years.

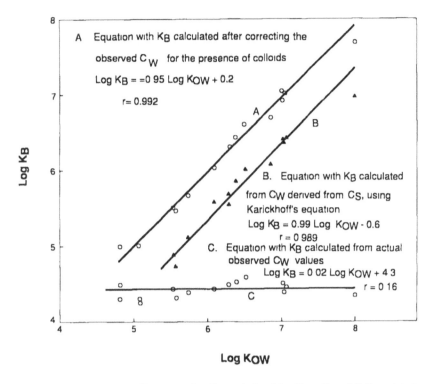

FIGURE 10. Plots of log K_B against log K_{OW} calculated by Connell et al.[83] from data in Oliver.[52] The C_B values used are the same for each plot, but C_W has been obtained by different means, as indicated.

Calculation of log K_{OW} values — In some investigations experimental values were not available for log K_{OW}, and so these values were calculated using various methods, some of which were previously outlined in Chapter 2. While in many cases quite accurate results can be expected from such calculations, there are many sources of error which lead to a degree of uncertainty regarding the accuracy of the results. Schüürmann and Klein[79] discuss some aspects of the use of calculated log K_{OW} values in bioconcentration investigations.

Inclusion of degradable compounds — Many of the investigations summarized in Table 3 include compounds which are subject to degradation by metabolic processes which occur to a significant extent within the organism after bioconcentration. The general characteristics of compounds which would be expected to bioconcentrate are discussed in Chapter 3. If significant degradation occurs after bioconcentration, the affected compounds may exhibit no bioconcentration at all, since the compound is effectively removed as bioconcentration proceeds, or bioconcentration occurs to a much lesser degree than expected. This effect is illustrated by some of the data in Table 3. For example, Davies and Dobbs[71] collated and critically evaluated a substantial body of bioconcentration data on lipophilic compounds. This included chlorohydrocarbons, phenols, phthalates, and several other types of organic compounds. With all of these substances together the relationship for "various organic" compounds shown in Table 3 (Reference 71, 1984) was obtained. This relationship has a positive intercept on the log K_B axis, or constant b value, rather than a negative intercept related to fat content. It also has a constant a value of 0.60 which is substantially different from the value of unity which would be expected if octanol was a perfect surrogate for fish lipid. However, if only chlorohydrocarbons are considered (see Table 3 and Reference 71, 1984) these constants have values in accord with theoretical expectations. Muir et al.[87] have investigated the bioconcentration of the readily hydrolyzable compounds, the triaryl phos-

TABLE 3
Characteristics of the Relationships Between Log K_B and Log K_{OW} and for Bioconcentration of Lipophilic Compounds by Fish

Constant a	Constant b	Number of values	r	Basis for K_B	Compound types	Ref.	Year
0.54	+0.12	8	0.95	Wet weight	Organic nonpolar	34	1974
1.16	−0.75	9	0.98	Wet weight	Various organic	66	1975
0.63	+0.73	11	0.79	Wet weight	—	36	1975
0.64	+0.73	11	0.79	—	—	35	1978
0.85	−0.70	55	0.95	Wet weight	Various organic	67	1979
0.94	−1.95	26	0.87	Wet weight	Various organic	35	1980
0.77	−0.97	36	0.76	Wet weight	Various organic	35	1980
0.46	+0.63	25	0.63	Wet weight	Various organic	68	1980
0.83	−1.71	8	0.98	Wet weight	Pesticides	69	1980
0.98	−0.06	6	0.99	Lipid wt.	—	70	1980
0.74	−0.77	40	—	—	—	71	1981
1.00	−1.32	36	0.97	Wet weight	Various organic	38	1982
1.02	−1.82	9	0.98	Wet weight	Phenols corrected for ionization	72	1982
1.02	−0.63	11	0.99	Wet weight	Chlorohydrocarbons	73	1983
0.79	−0.40	122	0.93	—	—	74	1983
0.94	−0.68	18	0.95	Wet weight	Chlorohydrocarbons	75	1984
0.60	+1.89	31	0.75	Wet weight	Various organic	71	1984
0.98	−1.36	20	0.90	Wet weight	Hydrocarbons and chlorohydrocarbons	71	1984
0.71	−0.92	17	0.98	Wet weight	Aromatic compounds	90	1984
1.09	−0.87	11	0.99	Wet weight	Chlorohydrocarbons	76	1985
0.96	−0.56	16	0.98	Wet weight	Chlorohydrocarbons	76	1985
0.89	+0.61	18	0.95	Lipid wt.	Chlorohydrocarbons	39[a]	1985
0.96	+0.25	18	0.96	Lipid wt.	Chlorohydrocarbons	39[a]	1985
0.61	+0.69	11	0.84	—	—	77	1985
0.94	−1.19	49	0.89	Wet weight	Various organic	78	1988
0.95	−1.06	30	0.99	Wet weight	Chlorohydrocarbons and PAH	78	1988
0.94	−1.00	33	0.85	Wet weight	Chlorohydrocarbons and related compounds	33	1988
0.75	−0.32	32	0.8	Wet weight	Various organic	79	1988
0.78	−0.35	22	0.95	Wet weight	Chlorohydrocarbons and PAH	79	1988

Note: Relationships between log K_B and log K_{OW} takes the form log K_B = a log K_{OW} + b.

[a] Triolein-to-water partition coefficient used.

phates, by fish. In this case the ease of hydrolysis, measured as the half life or rate constant for hydrolysis in alkaline acetone water solutions, was a better measure of bioconcentration than log K_{OW}. These compounds bioconcentrate consistently less than would be expected from Mackay's equation[38] (see Table 4). This is due to the hydrolytic removal of the bioconcentrated compounds from the fish, resulting in biotic concentrations less than would occur with a nonbiodegradable compound. It is suggested that QSARs based on log K_{OW} alone will apply most accurately to effectively nonbiodegradable compounds. The common chemical types which are least biodegradable are the chlorohydrocarbons and also the polyaromatic hydrocarbons. Other chemical types would be expected to deviate from the QSARs as described above, according to the amount of degradation which occurs; probably other factors are involved as well.

Nonattainment of equilibrium — All of the relationships developed for QSARs previously discussed depend on the attainment of equilibrium between organism and water. If equilibrium is not attained these relationships do not apply, and the actual observed rela-

TABLE 4
Measured and Calculated Log K_B Factors for Bioconcentration of Triaryl Phosphates by Fish

Compound	Log K_{OW}	Calculated log K_B[a]	Measured log K_B
tri-p-cresyl	4.62	3.30	2.89
			2.85
tri-m-cresyl	5.11	3.79	2.49
			2.66
triphenyl	5.11	3.79	2.51
			2.62
t-butylphenyl-diphenyl	5.12	3.80	2.89
			2.76

[a] Calculated using Mackay's equation.[38]

Data from Muir et al.[87]

tionships deviate from the expected values. This is discussed in more detail later in this chapter, Section III.A.

When considering the most appropriate relationships from those in Table 3, the factors outlined above should be kept in mind, together with the consideration that the development of relationships using a number of sets of data is usually more accurate than relationships from a single data set.

Connell and Hawker,[33] Davies and Dobbs,[71] Mackay,[38] Connell and Schüürmann,[78] and Schüürmann and Klein[79] have collated sets of data, and in some cases critically evaluated its accuracy and applied the data on specifically appropriate groups of compounds, e.g., chlorohydrocarbons and PAH. The relationships obtained had constant a values of 0.94, 0.98, 1.00, 0.95, and 0.78, respectively. Also, constant a values obtained for chlorohydrocarbons by Oliver and Niimi[73,76] were 1.02, 1.00, and 0.96. This suggests that constant a has a value close to unity, indicating that octanol is a good surrogate for fish lipid.

The values for constant b for this same set of relationships are -1.00, -1.36, -1.32, -1.06, and -0.35, respectively, with Oliver and Niimi[73,76] reporting values of -0.63, -0.87 and -0.56. These values correspond with fat percentage contents of 10, 4.3, 4.8, 8.8, 47, 23, 14, and 27, respectively. Kenaga and Goring[35] report that the fat content of fish ranges from about 1 to about 16%, with variations depending on such factors as age and species. This suggests that constant b will have values from -2 to -0.79 depending on the fish fat content. Thus, the equations of Connell and Hawker,[33] Davies and Dobbs,[71] Mackay,[38] and Connell and Schüürmann[78] are generally applicable to bioconcentration of lipophilic compounds by fish. Information on the fish lipid content is needed to apply these relationships in specific situations.

It should be remembered that the relationships outlined above apply to lipophilic compounds. Compounds having log K_{OW} values greater than approximately 6 to 7, and including the lipophilic compounds, are referred to here as hydrophobic compounds. The superhydrophobic compounds, i.e., log K_{OW} greater than 6 to 7, exhibit a falling log K_B with increasing log K_{OW}. The relationship of log K_B to log K_{OW} for hydrophobic compounds using fish as an example, has been discussed in this chapter, Section II.A.2.

Water solubility (S) has often been used in QSARs with fish bioconcentration. Previously, (this chapter Section II.A.1.) it was shown that

$$\log K_B = (3.25 a + b) - a \log C_l$$

where C_l was measured in moles m^{-3} and corrected using the melting point.

Using Mackay's values[38] (see Table 3) for constants a and b, then,

$$\log K_B = 1.93 - \log C_l$$

Table 5 contains a summary of the observed relationships between log K_B and log S. In this situation S is used to represent water solubility in any mass to volume units. Since constants a and b are altered by the units used to measure water solubility, these values will not in fact be consistent. Thus, the designations x and y are used to represent the values of constant a and constant (3.25 a + b), respectively. These data suggest that water solubility has considerable potential for QSAR, but further data are needed to firmly establish the nature of these relationships.

A variety of other molecular characteristics has been evaluated as QSAR for bioconcentration in fish, as summarized in Table 6. The soil-to-water partition coefficient expressed in terms of organic carbon (K_{oc}) was investigated around about 1980. Although useful, this characteristic has the disadvantage of not being measured under standardized conditions. For a satisfactory molecular characteristic in QSAR work, the characteristic must be capable of being measured under precisely defined conditions, allowing reproduction of results. With the K_{oc} value the type of soil used in measurement exhibits considerable variation in chemical components.

The molecular connectivity indices have shown reasonable correlations with log K_B (see Table 6). However, Brooke et al.[88] found a poor relationship between log K_B and the zero order connectivity index, using a wide array of 151 different types of chemicals. But when the chemicals considered are restricted to only chlorohydrocarbons, a quite close correlation is obtained. With this restricted range even such a simple index as molecular weight exhibited a close correlation, as illustrated in Figure 11.

Brooke et al.[88] fitted a parabola to their data, as did Sabljic and Protic[89] with bioconcentration data on compounds which were principally chlorohydrocarbons. The equation which exhibited the highest correlation found by Sablljic and Protic[89] was

$$\log K_B = -0.171\ (^2X^v) + 2.253(^2X^v) - 2.392 \qquad (r = 0.97)$$

where $^2X^v$ is the second order valence molecular connectivity index. This is in accord with the discussion on hydrophobic compounds in which parabolic and other polynomial expressions are used to explain the bioconcentration data.

Investigations by Connell and Shüürman[78] have suggested that molecular refraction, solvent accessible surface area, and solvent accessible volume are more satisfactory molecular descriptors to estimate bioconcentration with fish than the connectivity indices. It is interesting that the relationships reported by these authors exhibit an approximate parabolic shape, although the curvature is only clearly evident with the higher molecular weight compounds which do not extend into the superhydrophobic range to any great extent. Similar to the other investigations, these relationships apply best with the chlorohydrocarbons and the PAHs.

Somewhat similar results were obtained by Oliver[75] using the parachor. He found that this parameter can be a useful predictor of bioconcentration of lipophilic compounds with closely related groups such as the chlorobenzenes, but this parameter was less useful for a more chemically diverse group.

These latter molecular characteristics have an advantage in that they can be calculated from the molecular structure and thus do not require experimental work. This means that by the use of computer programs rapid calculations of these molecular characteristics, and subsequently the log K_B values, can be made.

TABLE 5

Characteristics of the Relationships Between Log K_B and Log S for Bioconcentration of Lipophilic Compounds by Fish

Value −x	Value y	Solubility units	Number of values	−r	Basis for K_B	Compound types	Ref.	Year
1.18	5.99	ppb	11	0.87	Wet wt.	Chlorohydrocarbons	80	1973
0.39	4.00	ppb	11	0.92	Wet wt.	Chlorohydrocarbons	82	1985
0.51	3.41	μM L^{-1}	8	0.93	Wet wt.	Various organic	81	1977
0.56	2.80	ppm	36	0.72	Wet wt.	Various organic	35	1980
0.63	2.18	ppm	50	0.66	Wet wt.	Various organic	35	1980
0.32	3.71	ppb	25	0.56	Wet wt.	Various organic	68	1980
0.55	2.83	ppm	42	—	—	—	Recalculated in 71	1981
0.44	4.36	μg L^{-1}	29	0.80	Wet wt.	Various organic	Recalculated in 71	1984
0.36	3.05	μM L^{-1}	29	0.82	Wet wt.	Various organic	71	1984

Note: Relationships between log K_B and log S take the form log K_B = y − x log S, where S is water solubility in mass to volume units.

Based on Davies and Dobbs[71] Table 1.

TABLE 6
Characteristics of the Relationships Between Log K_B and Various Molecular Characteristics for Bioconcentration of Lipophilic Compounds by Fish

Characteristic	Value x	Value y	Number of values	r	Compound types	Ref.	Year
Log K_{oc}	+1.12	−1.58	13	0.87	Various organic	35	1980
Log K_{oc}	+1.23	−2.02	22	0.91	Various organic	35	1980
Log K_{oc}	+0.70	−3.83	23	0.73	Various pesticides	84	1980
Log K_{oc}	+1.25	−1.22	—	—	Predicted	85	1981
1st Order valence molecular connectivity index	+0.79	+0.15	21	0.96	Various organic	86	1983
1st Order Randic Index	+0.58	+0.54	30	0.88	Chlorinated hydrocarbons and PAHs	78	1988
Molecular weight	9.93×10^{-3}	+1.00	30	0.81	Chlorinated hydrocarbons and PAHs	78	1988
Molecular refraction	5.54×10^{-2}	+0.19	30	0.91	Chlorinated hydrocarbons and PAHs	78	1988
Solvent accessible surface area	1.39×10^{-2}	−1.89	30	0.92	Chlorinated hydrocarbons and PAHs	78	1988
Solvent accessible molecular volume	6.94×10^{-3}	−0.77	30	0.91	Chlorinated hydrocarbons and PAHs	78	1988

Note: Relationships take the form $\log K_B = y + x$ (characteristic).

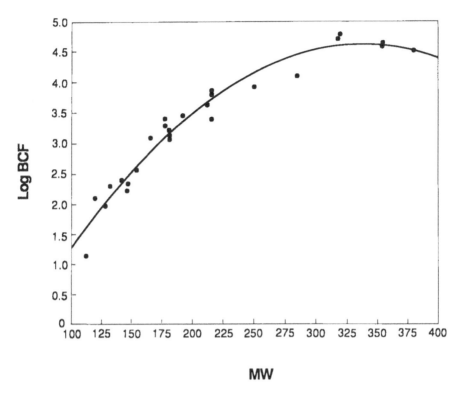

FIGURE 11.   Plot of the relationship between log K_B and molecular weight for chlorohydrocarbons. (From Brooke, D. N., Dobbs, A. J., and Williams, N., *Ecotoxicol. Environ. Saf.*, 11, 251, 1986. Copyright Academic Press. With permission.)

5. Molluscs

The information available on QSARs for bioconcentration of lipophilic compounds by molluscs is summarized in Tables 7 and 8. In these relationships only K_{OW} and S have been used. The comments regarding inclusion of inaccurate data, biodegradable compounds, and nonattainment of equilibrium made in relationship to QSARs for fish are generally applicable to these data also.

While most of these data are derived from laboratory experiments, some field data are also included. For example, the information on mussels and oysters in Table 7 reported by Ogata et al.[90] is derived from field information. However, the experiments reported by Ogata et al.[90] were carried out on alkyl dibenzothiophenes. This limited range of chemical type and their possible susceptibility to biodegradation raises questions as to the general applicability of these expressions.

Ernst[12] has reported that the lipid content of mussels ranges from 1.2 to 1.8%, but that there can be considerable seasonal and other variations. Based on the discussion in this chapter, Section II.A.1., if octanol is a good surrogate for mollusc lipid, then a constant b value for the log K_B to log K_{OW} relationship would be about -1.80 with a constant a of unity. The relationships reported by Geyer et al.[11] and Hawker and Connell[61] (Table 7) are in general agreement, which is in approximate accord with these values for constant a and constant b. Since the equation reported by Hawker and Connell[61] has been based on more recent and probably more reliable data, this equation may be more generally applicable.

The data on the log K_B to log S relationship (Table 8) is limited. However, the reasonably close agreement between the two reported equations, together with the high correlation coefficients suggests these equations may be accurate and generally applicable.

TABLE 7
Characteristics of the Relationships Between Log K_B and Log K_{OW} for Bioconcentration of Lipophilic Compounds by Molluscs

Constant a	Constant b	Number of values	r	Basis for K_B	Compound types	Organism	Ref.	Year
0.86	−0.81	16	0.96	Wet wt.	Various organic	Mussel	11	1982
0.16	1.52	14	0.71	Wet wt.	Alkyl dibenzothiophenes	Short-necked clam	90	1984
0.49	1.03	14	0.62	Wet wt.	Alkyl dibenzothiophenes	Oyster	90	1984
0.31	1.63	14	0.64	Wet wt.	Alkyl dibenzothiophenes	Mussel	90	1984
0.84	−1.23	34	0.83	Wet wt.	Mainly chlorohydrocarbons	Mollusc	61	1986

Note: Relationships take the form log K_B = a log K_{OW} + b.

Based on Connell.[20]

TABLE 8
Characteristics of the Relationships Between Log K_B and Log S for Bioconcentration of Lipophilic Compounds by Molluscs

Value −x	Value y	Solubility units	Number of values	−r	Basis of K_B	Compound types	Ref.	Year
0.84	5.15	$\mu g\ L^{-1}$	7	0.96	Wet wt.	Chlorohydrocarbons	12	1977
0.68	4.94	$\mu g\ L^{-1}$	16	0.94	Wet wt.	Various organic	11	1982

Note: Log K_B = y − x log S, where S is water solubility.

Based on Connell.[20]

Despite the discussion and results above, a degree of caution is necessary in applying these relationships. Molluscs comprise a relatively diverse group of biota within which individual species may have different bioconcentration characteristics. For example, Hawker and Connell[61] report that the hard clam *(Mercenaria mercenaria)* exhibits different bioconcentration characteristics from the soft clam *(Mya arenaria)*. This may be due to different biodegradation capacities of the two species.

6. Sediment-to-Water Systems

QSARs for lipophilic compounds distributed in sediment-to-water equilibrium systems have been extensively investigated. A summary of some of the reported relationships is shown in Table 9. The sediment-to-water system is briefly reviewed here for two reasons. First, this system is basically similar in many respects to bioconcentration, while exhibiting several distinctive differences. Second, the sediment-to-water system has a major impact on the availability of xenobiotic lipophilic compounds to aquatic organisms. The accumulation of compounds in ambient water by sediments is characterized by the water-to-sediment partition coefficient at equilibrium, which is usually expressed in terms of organic carbon as

$$K_{oc} = C_s/(C_w\, f_{oc})$$

where C_s is the sediment concentration, f_{oc} is the fraction of organic carbon in the sediment, and C_w is the water concentration.

This phenomenon can be interpreted in a somewhat similar manner to bioconcentration by aquatic organisms if only nondegradable neutral lipophilic compounds are considered. If octanol is a perfect surrogate for the organic carbon in the sediment, then

$$K_{oc} = K_{ow}$$

But by analogy with organisms in which only lipid is active, only a fraction of the organic carbon (F_A) may be active in the partition process. Now a more accurate representation of K_{oc} would be

$$K_{oc} = F_A\, K_{ow}$$

taking logarithms

$$\log K_{oc} = \log K_{ow} + \log F_A \qquad (16)$$

This is analogous to the expression for bioconcentration found to be applicable to aquatic organisms, i.e.,

$$\log K_B = \log K_{ow} + \log y_L$$

However, with sediments a much greater variation in the composition of the lipophilic compound sorbing fraction in the organic carbon could be expected. The literature also indicates that the sorption of lipophilic compounds by sediments involves more complex processes than simple solubilization in lipid. But for nondegradable, neutral, lipophilic compounds which are sorbed by processes similar to lipid and octanol sorption, Equation 16 may be applicable.

Equations for the relationship between $\log K_{oc}$, $\log K_{ow}$ are shown in Table 9 and take the general form

TABLE 9
Characteristics of Some of the Relationships Between Log K_{oc} and Various Parameters for Lipophilic Compounds in a Sediment-to-Water System

Characteristic	Value x	Value y	Number of values	r	Compound types	Ref.	Year
Log K_{OW}	0.72	+0.49	13	0.95	Mainly methylated and halogenated benzenes	91	1981
Log K_{OW}	1.03	−0.18	13	0.91	Pesticides	91	1982
Log K_{OW}	0.99	−0.35	5	1.00	Polyaromatic hydrocarbons	91	1981
Log K_{OW}	1.00	−0.32	22	0.98	Polyaromatic hydrocarbons	91	1980
Log K_{OW}	1.00	−0.21	10	1.00	Polyaromatic hydrocarbons	94	1979
Log K_{OW}	0.54	1.38	45	0.74	Agricultural chemicals	94	1980
Log K_{OW}	0.94	−0.01	19	0.95	Triazines, nitroanilines	94	1981
Log K_{OW}	0.52	0.64	30	0.84	Phenyl areas and alkyl-n-phenyl carbonates	94	1981
Log S	−0.56	4.04	15	0.99	Halogenated hydrocarbons	92	1979
1st Order molecular connectivity index	0.55	0.45	37	0.97	Hydrocarbons and chlorohydrocarbons	93	1981

Note: Log K_{oc} = x (characteristic) + y.

$$\log K_{oc} = x \log K_{OW} + y$$

The discussion above suggests that in appropriate situations, where octanol is a good surrogate for the lipophilic compound sorbing fraction in organic carbon, the constant x should be about unity. It also indicates that constant y should always be negative and be related to the proportion of organic carbon which is active in sorbing lipophilic compounds.

On consulting Table 9 the compounds likely to most closely fit the criteria above are methylated and halogenated benzenes, PAHs, chlorobenzenes, and alkyl benzenes. The relationships applicable to these groups have constant x values of 0.72, 0.99, 1.00, and 1.00, respectively, reading down Table 9. Similarly, constant y values are $+0.49$, -0.35, -0.32, and -0.21. These latter values would represent about 50 to 60% of lipid sorbing material in the organic carbon when the value of $+0.49$ is excluded. Most of the other compounds in the table contain complex functional groups which may lead to divergences from the expected relationships.

Such an approach may have application in sediment-to-water QSARs under the circumstances outlined above. However, it should be emphasized that this interpretation is speculative and lacks a clear set of data to demonstrate its applicability.

In many situations, sediment-to-water systems have proportions of dissolved organic matter and colloids in the water phase. The colloidal and dissolved organic matter fraction contains relatively high concentration of sorbed lipophilic compounds, and if a water sample is taken and analyzed, it will contain relatively high concentrations of lipophilic compounds. Farrington and Westall[94] have discussed this phenomenon and illustrated how it can have a major effect on $\log K_{oc}$ to $\log K_{OW}$ relationships. The major deviations occur with compounds having comparatively high $\log K_{OW}$ values and lead to a curvilinear plot. This occurs most markedly in field situations where large amounts of colloidal and organic matter are often present in the ambient water.

7. Overall Aspects of QSARs for Bioconcentration of Lipophilic Compounds

In the previous sections the QSARs reported for the various biota were evaluated. Suggestions were made as to which of these relationships are most likely to be accurate and applicable; these are summarized in Table 10. In addition, a representative set of these relationships is presented in Figure 12.

If octanol perfectly represents the organism lipid, then

$$K_{BL} = K_{OW}$$

and

$$\log K_{BL} = 1 \log K_{OW} + \log 1 = 1 \log K_{OW} + O$$

where K_{BL} is the bioconcentration factor expressed in terms of lipid weight.

This equation is represented in Figure 12 with a slope of unity and a $\log K_B$ intercept of 0. If K_B is expressed in wet weight terms, then parallel lines are obtained with intercepts related to the lipid content of the organism, as shown in Figure 4. These relationships were previously discussed in this chapter, Section II.A.1. This type of relationship has been shown to exist for lipophilic compounds (i.e., $\log K_{OW}$ about 2 to about 6.5) as illustrated by the results in Table 10 and Figure 12. Superhydrophobic compounds, i.e., $\log K_{OW} >$ about 6.5, exhibit different relationships, as described in this chapter, Section II.A.2.

The most applicable set of constant a values is generally in reasonably close agreement for different biota and for the sediment-to-water system. In addition, most of these values are close to unity. This suggests that generally, octanol is a good surrogate for organism

TABLE 10
Characteristics of the Most Applicable Relationships Between Log K_B and Log K_{OW} for Various Biota and the Sediment-to-Water System

Biota	Constant a	Constant b	Lipid equivalent to constants b (%)	Actual lipid (%)	Range of log K_{OW}	Ref.
Microorganisms	0.91	−0.36	44 (dry wt.)	—	3—7	2
Daphnids	0.90	−1.32	4 (wet wt.)	—	2—8	61
Poly- and oligochaetes	0.99	−0.60	25 (dry wt.)	—	4—8	83
Fish	0.94	−1.00	10 (wet wt.)	1—16 (wet wt.)	3—6	33
Fish	0.98	−1.36	4.3 (wet wt.)	1—16 (wet wt.)	1.5—6.5	71
Fish	1.00	−1.32	4.8 (wet wt.)	1—16 (wet wt.)	0.5—6.0	38
Fish	0.95	−1.06	8.8 (wet wt.)	1—16 (wet wt.)	2—6	78
Molluscs	0.84	−1.23	5.9 (wet wt.)	1.2—1.8 (wet wt.)	3.5—8	61
Sediment water system	0.99	−0.35	—	—	—	91
Sediment water system	1.00	−0.32	—	—	—	91
Sediment water system	1.00	−0.21	—	—	2—6	94
Sediment water system	0.72	+0.49	—	—	3—5	94

Note: Log K_B (or Koc) = a log K_{OW} + b.

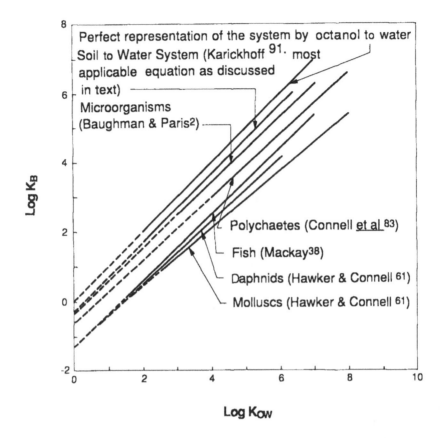

FIGURE 12. Relationship of log K_B to log K_{ow} for different groups of biota, the sediment-to-water system perfect, and the system which is represented by the octanol-to-water system. The solid line gives the range of values used in obtaining the relationship, with the broken line representing extrapolation to the log K_B axis.

lipid. Also, it may reasonably represent the fraction of organic carbon in sediments which sorbs nonpolar nondegradable xenobiotic compounds. In fact, Sondergren[114] carried out comparison experiments between hexane-filled bags and aquatic organisms, and has suggested that this bag system is similar in bioconcentration behavior to biota.

The constant b values are generally in accord with the lipid content of the organisms where this is known. This is a more difficult factor to evaluate since it may vary substantially within a given group, or within a given species, according to seasonal and other factors which may cause a variation in lipid content. But this analysis also generally indicates that octanol is a good surrogate for biota lipid.

Overall, this data indicates that octanol is a good surrogate for organism lipid and that K_{OW} is a good predictor of K_B. It suggests that if K_B values were expressed in terms of biota lipid weight, all nondegradable lipophilic compounds would fall on the perfect representation line. Of course, stereochemical, kinetic, and other factors may cause deviation from this, as discussed in this chapter, Section III. In addition, different biota have different capacities to degrade the various hydrophobic compounds, and this would be expected to influence the K_B values as well.

ORGANISM

FIGURE 13.   Single-compartment model for the uptake and clearance of a lipophilic chemical by an organism.

III. INFLUENCE OF VARIOUS FACTORS ON QSARS FOR BIOCONCENTRATION

A. KINETICS OF BIOCONCENTRATION

1. General Characteristics of the Single Compartment Model

Theoretical treatment of the kinetics of chemical transfer in organisms was developed in pharmacology by Atkins[95] and applied to the uptake and clearance of environmental contaminants by Moriarty.[96] The most useful model for the kinetics of bioconcentration is based on the single compartment approach utilizing a single compartment to represent an organism. Moriarty[96] has also used a two compartment system, with one compartment to represent a peripheral system and the other a central system. Applied to available data, he found that this gave a better explanation, but the increase in complexity makes this approach more difficult to apply effectively. Walker[97] has also reported that for some fish and pollutant combinations, the first order approach has failed to provide useful results.

Bioconcentration can be seen as the balance of uptake and clearance processes, as shown in Figure 13. These processes proceed by first order kinetics characterized by the rate constants k_1 and k_2, respectively. Clearance is a physical process due to the reverse movement of molecules as a result of the concentration of compound in the organism. It does not include or result from biodegradation of the compound. Then, if the compound involved is lipophilic and nondegradable, the rate of change of its concentration in biota is expressed by

$$\text{rate of change in biotic concentration} = \text{rate of uptake} - \text{rate of clearance}$$

this means that

$$dC_B/dt = k_1 C_W - k_2 C_B \tag{17}$$

where C_B is concentration in the biota and C_W concentration in water. Since the amount of compound in the water represents a large reservoir compared to the relatively low amount that can be taken up by biota, C_W can be regarded as constant in any particular situation. By integration and rearrangement of the equation above

$$C_B = (k_1/k_2) C_W (1 - e^{-k_2 t}) \tag{18}$$

This predicts that C_B will exhibit a steady increase in concentration with time, but with a declining rate of increase, as shown in Figure 14. Thus, t continues to increase until $e^{-k_2 t}$ is zero and the C_B curve is parallel to the time axis. At this time

$$C_B = (k_1/k_2) C_W$$

and

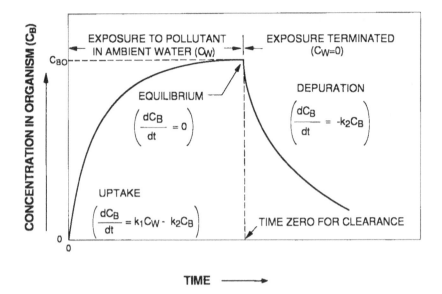

FIGURE 14. Uptake and clearance patterns of a lipophilic pollutant over time with an organism as represented by first order kinetics.

$$C_B/C_W = k_1/k_2 = K_B \qquad (19)$$

Thus, K_B values can be established from kinetic data. This can also be seen as the stage where rates of uptake and clearance are equal and

$$dC_B/dt = O = k_1 C_W - k_2 C_B$$

then

$$k_1 C_W = k_2 C_B$$

Since the log K_B-to-log K_{OW} relationship requires that equilibrium be established, it is important to evaluate the time period needed to reach this stage. The theoretical time period to reach equilibrium occurs when $e^{-k_2 t}$ is zero, and thus when t is infinity. But effective equilibrium can be considered to be reached at t_{eq}, when C_B is 0.99 of the C_B value at infinity. Thus from Equation 18

$$C_B \text{ (at effective equilibrium)} = (k_1/k_2) C_W (1 - e_{eq}^{-k_2 t})$$
$$= 0.99 (k_1/k_2) C_W$$

thus

$$0.99 = 1 - e_{eq}^{-k_2 t}$$

and

$$t_{eq} = 4.605 (1/k_2) \text{ or } \log t_{eq} = \log (1/k_2) + 0.663 \qquad (20)$$

If exposure to the compound is terminated, for example, by transfer to uncontaminated water, then $C_W = 0$, and so $k_1 C_W = 0$, and

$$dC_B/dt = -k_2 C_B$$

Thus, before, during exposure, both uptake and clearance were operating, but now, in uncontaminated water, uptake does not occur and only clearance is in operation. By integration and rearrangement

$$C_B = C_{BO} e^{-k_2 t}$$

and

$$\ln C_B = \ln C_{BO} - k_2 t$$

where C_{BO} is the initial concentration at time zero for the clearance period.

This shows that as t increases C_B declines, but the rate of decline decreases with increasing time (see Figure 14). Also, since C_{BO} and k_2 are constants, $\ln C_B$ is linearly related to time. When half of the initial compound has been cleared, then $C_B = C_{BO}/2$ and the half life, $t_{1/2}$, is represented by

$$t_{1/2} = (\ln 2)/k_2$$

The persistence of a compound can be characterized as a half life, and is due to the physical loss of compound and not by biodegradation and excretion of the degradation products. If these processes are significant, then a different half life, most likely considerably shorter, would be operating.

2. Kinetic Rate Constants and Time to Equilibrium for Lipophilic Compounds

With many aquatic organisms, bioconcentration directly from water occurs through the gills, or other respiratory surfaces, and has been shown to follow first order kinetics. Thus, the rate of uptake is proportional to the concentration in the water, C_W, and

$$dC_v/dt = k_1 C_W$$

where C_v is the concentration in the organism due to uptake without allowing for the loss of compound due to clearance. Spacie and Hamelink[98] report that for a given compound, the k_1 value remains constant over a wide range of concentrations, provided the material remains in true solution. Neely[99] has viewed uptake as the partitioning of a compound in water at the gill surface. Uptake continues since there is fresh compound available as it comes in contact with the gill surface due to movement of water as a result of respiration. Thus

$$k_1 = ER_v/F$$

where E is the extraction or transfer efficiency across the gill membrane, R_v is the ventilation rate of respired water, and F is the weight of the fish.

The kinetic processes can also be evaluated by considering the log K_B-to-log K_{ow} relationship. Previously, this relationship for nonbiodegradable lipophilic compounds was found to have the following general form:

$$\log K_B = a \log K_{ow} + b$$

since

$$k_1/k_2 = K_B$$

then

$$\log (k_1/k_2) = a \log K_{OW} + b$$

This means that the ratio of the uptake and clearance rate constants is proportional to the log K_{OW} value, and suggests that there may be a relationship between the rate constants, individually, and log K_{OW}. Neely[99] in 1979 suggested a relationship between k_1, k_2, and K_{OW}, or S, for fish. A summary of the empirical relationships later developed between the rate constants (log k_1 and k_2) and log K_{OW} is shown in Table 11. The relationships between log $(1/k_2)$ and log K_{OW} take the general form

$$\log (1/k_2) = x \log K_{OW} + y \qquad (21)$$

where x and y are empirical constants.

It can be shown, using this equation and Equation 1, that

$$\log k_1 = (a - x) \log K_{OW} + (b - y))$$

However, it is important to note that these relationships have been established over the range of log K_{OW} values from about 2 to about 6.5 and may not be applicable outside this. In fact, Gobas et al.[103] have suggested that with log k_1, deviations from linearity commence at log K_{OW} values of less than 3 to 4, and log $(1/k_2)$ may be independent of log K_{OW} at low to moderate values, although there is a linear relationship at higher values.

An expression relating log t_{eq} and log K_{OW} can be obtained by substituting log $(1/k_2)$ from Equation 21 into Equation 20. Thus

$$\log t_{eq} = x \log K_{OW} + (y + 0.663)$$

The actual equations found by Hawker and Connell[61,102] for daphnids, molluscs, and fish were in the general form above, where t_{eq} was measured in hours and constant x was 0.507, 0.540, and 0.663, and constant (y + 0.663) was −1.390, −0.320, and −0.284, respectively. A plot of these relationships is shown in Figure 15 and indicates the extended time periods required to establish equilibrium for compounds of relatively high log K_{OW} values. Although Hawker and Connell[61,102] extrapolated beyond log K_{OW} values of 6.5 to obtain equilibrium times, these can only be regarded as speculative; in later work, outlined in this chapter, Section III.A.3, a more accurate interpretation was made.

Later these authors[104] refined this set of relationships for bioconcentration of lipophilic compounds in the log K_{OW} range from 2 to about 6.5. This refinement was based on the use of direct relationships of the rate constants (k_1 and k_2) to K_{OW}, rather than the logarithmic values. The available data yields significant relationships for this data in both forms and does not allow a distinction to be made as to which is, in fact, most applicable. The equations obtained in this way were

$$1/k_2 = 1.42 \times 10^{-3} K_{OW} + 12.01$$

$$k_1 = (0.048 K_{OW})/(1.42 \times 10^{-3} K_{OW} + 12.01)$$

TABLE 11
Characteristics of the Constants in the Relationships Between the Rate Constants in Bioconcentration and Log K_{OW}

Biota	Rate constant	Value x or (a − x)	Value y or (b − y)	Units (ml g⁻¹·day⁻¹)	Number of values	r	Compound types	Range of log K_{OW}	Ref.
Fish	k_1	0.18	1.98	day⁻¹	13	0.74	Chlorohydrocarbons	2.5–8	98
Fish	k_2	−0.41	1.47	day⁻¹	13	0.95	Chlorohydrocarbons	2.5–8	98
Fish	k_1	0.46	1.84	day⁻¹	5	0.9	Chlorobenzenes	3.5–6	101
Fish	k_2	−0.42	1.44	day⁻¹	5	0.99	Chlorobenzenes	3.5–6	101
Fish	k_1	0.34	−0.37	h⁻¹	21	0.78	Mainly chlorobenzenes	2–7	102
Fish	$1/k_2$	0.66	−0.95	h	21	0.97	Mainly chlorobenzenes	2–7	102
Molluscs	k_1	0.30	−0.25	h⁻¹	34	—	Mainly chlorobenzenes	3–8	61
Molluscs	$1/k_2$	0.54	−0.98	h	34	0.91	Mainly chlorobenzenes	3–8	61
Daphnids	k_1	0.39	+0.74	h⁻¹	22	—	Various organic	2–6	61
Daphnids	$1/k_2$	0.51	−2.05	h	22	0.98	Various organic	2–6	61

Note: Log $(1/k_2)$ or $k_2 = x$ log $K_{OW} + y$ and log $K_1 = (a − x)$ log $K_{OW} + (b − y)$.

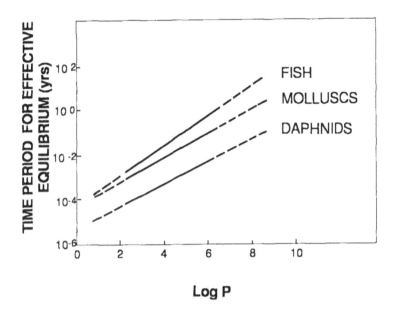

FIGURE 15. The relationship between the logarithm of the time period to establish effective equilibrium and log K_{ow} for various organisms. (From Hawker, D. W. and Connell, D. W., *Ecotoxicol. Environ. Saf.*, 11, 184, 1986. Copyright Academic Press. With permission.)

$$t_{eq} = 6.54 \times 10^{-3} K_{OW} + 55.31$$

These equations give somewhat similar results to those previously described, which were based on the logarithmic relationships in the log K_{OW} range from 2 to about 6.5. However, generally they take a curvilinear form with k_1 and K_B reaching maximum values.

3. Kinetic Rate Constants and Time to Equilibrium for Hydrophobic Compounds

In consideration of QSARs for bioconcentration there were two sets of relationships discussed. There were those for the lipophilic compounds in the log K_{OW} range from 2 to about 6.5, and those for the hydrophobic compounds which include the lipophilic compounds, as well as those with values beyond this log K_{OW} range. Similarly, in considering the kinetic relationships, a distinction can be made between these two groups; with the wider range of hydrophobic compounds, different overall relationships are apparent. Unfortunately, with hydrophobic compounds little data are available on compounds with log K_{OW} values greater than about 6.5. However, a limited amount is available on fish, and only this group of biota is considered in this section.

Connell and Hawker[33] found in investigating the kinetic data for hydrophobic compounds that the following relationships apply. These are based on polynomial expressions derived from data on chlorohydrocarbons and related compounds over the range of log K_{OW} values from 2.5 to 9.5.

$$\log k_1 = 2.92 - 9.86 \times 10^{-2}(\log K_{OW} - 5.87)^2$$

$$\log (1/k_2) = 6.9 \times 10^{-3}(\log K_{OW})^4 - 1.85 \times 10^{-1}(\log K_{OW})^3$$
$$+ 1.65(\log K_{OW})^2 - 5.34 \log K_{OW} + 5.27$$

$$\log t_{eq} = 6.9 \times 10^{-3}(\log K_{OW})^4 - 1.85 \times 10^{-1}(\log K_{OW})^3$$
$$+ 1.65(\log K_{OW})^2 - 5.34 \log K_{OW} + 5.93$$

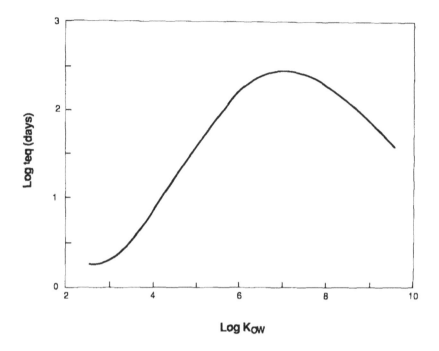

FIGURE 16.   The relationship of time period to establish equilibrium (log t_{eq}) to log K_{OW} for bioconcentration of hydrophobic compounds by fish. (Adapted from Connell, D. W. and Hawker, D. W., *Ecotoxicol. Environ. Saf.*, in press. Copyright Academic Press)

The curves represented by these equations are either parabolas or a similar shape, over the log K_{OW} range from 2.5 to 9.5. Over the lipophilic range of low K_{OW} from 2 to about 6.5, the curves are approximately linear, and follow the linear relationships as described in the previous section on lipophilic compounds.

The relationships of log t_{eq} to log K_{OW} is of particular importance in understanding QSARs for bioconcentration. This relationship is shown graphically in Figure 16. This shows that the time to establish equilibrium increases with log K_{OW} to a maximum at log K_{OW} of approximately 7 when the time will be 325 days or 0.89 years. If the exposure time periods are less than these periods, then the log K_B value will be less. Nonequilibrium log K_B values can be calculated from the following equation derived from Equation 18:

$$C_B = (k_1/k_2)\, C_W\, (1 - e^{-k_2 t}) \tag{18}$$

Thus

$$K_B = k_1/k_2\, (1 - e^{-k_2 t})$$

and

$$K_{B(exptl)} = K_{B(infinity)}\, (1 - e^{-k_2 t})$$

where $K_{B(exptl)}$ is the nonequilibrium bioconcentration factor after an exposure period of t, and $K_{B(infinity)}$ is the bioconcentration factor at equilibrium.

The effect of exposure time period on log K_B values, calculated as above, is shown in Figure 17. This figure illustrates that the log K_B-to-log K_{OW} relationship can deviate from a straight line even in the log K_{OW} range for lipophilic compounds, depending on the exposure

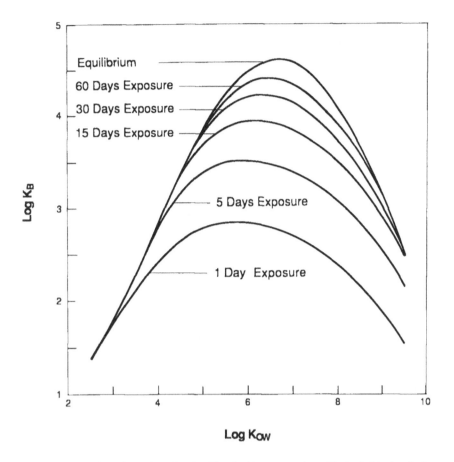

FIGURE 17. Plots of log K_B against log K_{OW} for bioconcentration of hydrophobic chemicals by fish at different exposure time periods. (Adapted from Connell, D. W. and Hawker, D. W., *Ecotoxicol. Environ. Saf.*, in press. Copyright Academic Press.)

period. In addition, the slope of the regression line and intercept on the log K_B axis will vary with exposure period. As the time period approaches the equilibrium exposure period, the values of the slope and intercept will approach those described for QSARs with different biota previously described. At equilibrium, an approximately parabolic shape is attained. But log K_B values for compounds having log K_{OW} greater than 6.5 are difficult to measure, because as the log K_B are declining, the biota concentrations and corresponding water concentrations are extremely low.

B. MOLECULAR STEREOCHEMISTRY

Several authors have noted differential bioaccumulation of different PCB isomers, as outlined by Shaw and Connell.[105] Many of these changes can be explained by differential metabolism of these different isomers.[94] Some substitution patterns on the biphenyl rings are resistant to degradation, while others are susceptible. In addition, molecular stereochemistry has been used to explain these differential bioconcentration patterns.

The PCB molecule exhibits substantially different stereochemistry depending on the chlorine substitution patterns on the two phenyl rings. Shaw and Connell[105] have developed an empirical steric effect coefficient (SEC) which they used to adjust the bioconcentration factor of different PCB congeners. The product of the SEC and log K_{OW} provided a satisfactory relationship with the observed bioconcentration factor measured with fish and polychaetes. It was suggested that the larger and more bulky molecules were bioconcentrated to a lesser extent than others with a more compact structure.

More recently, Opperhuizen et al.[106] have reported on a detailed investigation of the role of molecular size and shape in bioconcentration with a range of different types of compounds. This was outlined in Chapter 3. Molecules with a relatively large molecular cross section were subject to limited bioconcentration, or were essentially not bioconcentrated at all.

C. BIOLOGICAL AND ENVIRONMENTAL CONDITIONS
1. Size and Type of Organism

Previously, it was noted that k_1, the uptake rate constant, could be derived from

$$k_1 = ER_v/F$$

where E is the transfer efficiency of compounds across the gill membrane, R_v is the ventilation rate, and F is the weight of the fish.

R_v is a function of body weight, but generally it has been found that metabolic rate and ventilation rate per weight unit decline slightly with increasing weight.[100] Since weight is often related to age within a specific group of organisms, it could be expected that in these cases k_1 would decline slightly in proportion to age. Similarly, in general, k_1 should decline slightly with increasing size of fish. In all cases, related change would be expected in k_2. It would be expected, on a similar basis, that as a general rule small organisms in comparison to large organisms will have relatively faster uptake and clearance rates.

From the previous discussion on QSARs it was indicated that the intercept on the log K_B axis was related to organism lipid content. But it was also suggested that bioconcentration on a lipid basis could be directly related to log K_{OW} according to a relationship common to all organisms, i.e., $K_{OW} = K_{BL}$, where K_{BL} is the bioconcentration factor on a lipid basis. This relationship would probably be influenced by two factors. The first factor is the different metabolic capacity of different organisms. If an organism has a significant capacity to degrade certain types of compound, that organism will have little capacity to bioconcentrate those compounds. Second, the uptake and clearance rates and rate constants will be related to the rate of respiration of the type of organism. Thus, the apparent log K_B to K_{OW} relationship may differ after similar exposure times for different types of organisms, and will only be comparable after equilibrium has been achieved. In a number of field investigations on fish an increase in the concentration of xenobiotic chemicals has been observed at length. Connell[113] was able to explain this phenomenon in the striped bass as due to the slow rate of bioconcentration of the PCBs involved.

2. Route of Uptake of Xenobiotic Compounds

In this chapter, Section I.A., the major routes of uptake were described. Briefly, an organism can acquire persistent xenobiotic residues from the ambient water or from food. It was suggested that irrespective of which uptake route is involved, release to, and equilibrium with, the ambient water will occur through the respiratory surfaces. However, uptake by food, which would occur at the same time as uptake from ambient water, would alter the rates of uptake and clearance by the organism. Thus, observed QSARs at different time periods could be influenced and altered by these factors, although at equilibrium, all aquatic organisms would be expected to attain concentrations consistent with bioconcentration, as discussed in Chapter 7, Section III.C.3.

3. Biodegradation

The single compartment model for bioconcentration has been described in this chapter, Section III.A. In this treatment the clearance rate constant is due to the reverse process to uptake, which occurs as a result of the accumulation of chemical in the organism. If degradation occurs it results in an increase in the clearance of the compound over and above that due to the physical process outlined above. The rate constant for this process (k_3) is a more complex factor, which may possibly vary with time. This means it may not be constant, although further information is needed to provide an accurate description of this value. However, for the purposes of this discussion, it will be assumed that k_3 is constant.

The kinetics rate constants, previously described by Equation 19 without degradation, were in the following relationship at equilibrium:

$$C_B/C_W = k_1/k_2 = K_B$$

It can be shown that if the clearance rate constant is increased by k_3 due to degradation then at equilibrium

$$C_B/C_W = k_1/(k_2 + k_3) = K_B \text{ (deg)}$$

where K_B (deg) is the bioconcentration factor if degradation occurs. The k_3 value can have a wide range, depending on the susceptibility of the compound to degradation. But in all cases where k_3 is significant, K_B (deg) is less than K_B. Thus, biodegradation will reduce the bioconcentration factor in all situations. The amount of bioconcentration which occurs and the bioconcentration factor will depend on the amount of biodegradation which results. But if k_1 is less than $(k_2 + k_3)$ and this is much greater than k_2, then no bioconcentration at all will occur.

As a general rule, the chlorohydrocarbons exhibit no significant degradation over the period of bioconcentration, and often PAHs exhibit similar resistance to biodegradation. Thus these compounds usually exhibit the bioconcentration behavior expected from the log K_B to log K_{OW} QSAR. Other compounds usually exhibit varying degrees of biodegradation and thus have related degrees of deviation from this QSAR.

4. Physical and Chemical Conditions in the Ambient Water and Sediments

Temperature would be expected to affect the uptake rate constant by influencing the metabolic rate, and thus the respiration and ventilation rate. The magnitude and nature of this effect is related to the optimum and ambient temperatures which exist at the time. Spigarelli et al.[107] in experiments on the brown trout found that ambient temperature affected PCB bioaccumulation by affecting food consumption, growth, and lipid content. Somewhat similarly, the presence of chemicals in the environment that have toxic or other physiological effects may influence the ventilation rate and consequently, the uptake. Connell and Miller[108] have noted that low dissolved oxygen concentrations can increase the ventilation rate of aquatic organisms, which would also probably result in increased uptake rates.

Ionized compounds in the water mass are usually affected by pH and the presence of inorganic salts that alter the proportion of unionized compound present. This changes the observed K_{OW} value which would alter the K_B value of a given compound.[109] However, with unionized compounds this phenomenon would not be expected to operate, and in support of this Zaroogian et al.[77] found that marine species have similar bioconcentration characteristics to freshwater species. On the other hand, Murphy[110] found that a single species tolerant to different salinities bioconcentrated DDT and related compounds differently at different salinities.

The sediments in an aquatic area can be an important source of compounds to maintain water concentrations, and thus, organism concentrations. The presence of organic matter in sediments is a major factor influencing the significance of the sediments as a xenobiotic compound reservoir. Since high organic matter concentrations displace the sediment-to-water partition coefficient in favor of the sediments, the water concentrations are correspondingly lowered. Shaw and Connell,[111] as a result of field and laboratory work, concluded that the occurrence of petroleum hydrocarbons in sediments raised the partition coefficient between sediments and water, thereby decreasing the quantity of PCB available to poly-chaetes.

In many field situations there are significant quantities of colloidal matter and dissolved organic matter present in the ambient water. Thus, unless precautions are taken to remove or account for compounds sorbed to colloid and organic matter, the dissolved water con-centrations will be apparently higher than they are in fact. Landrum et al.[112] has described how this influences the apparent kinetics and bioconcentration factors. These authors have suggested that the reduced log K_B factors for compounds with log K_{OW} greater than about 6 may be due to this effect. At log K_{OW} values greater than about 6, a large proportion of the compound present in the ambient water is sorbed to the dissolved organic carbon and colloidal matter present. This effect increases with increasing low K_{OW}, so calculations of log K_B values exhibit a decline due to elevated values for the water concentrations.

5. Comparison of Laboratory and Field Data

Most of the results described in this chapter are based on laboratory experiments, although some field data are included also. Direct comparisons between field and laboratory data present a number of difficulties. The period of exposure of an organism can be for the full life cycle. It includes different life stages with grossly different sizes and physiology, seasonal or randomly periodic. Concentrations may show considerable variability throughout the range of movement of an organism, and thus concentration and period of exposure may be very difficult to determine. The presence of other chemicals, which may exercise a physiological or physical effect on bioconcentration, also needs to be known and the effect on biocon-centration evaluated.

Farrington and Westall[94] outline some comparisons of field data with QSARs for log K_B to log K_{OW} and describe some of the difficulties involved. Davies and Dobbs[71] reviewed the relationship between laboratory and field data and concluded that the data were too restricted to arrive at a conclusive result. However, they suggested that the limited information available indicates that log K_B vs. log K_{OW} QSAR reflects bioconcentration in polluted fresh water.

REFERENCES

1. **Hansch, C. and Fugita, T.,** ρ-σ-π analysis. A method for the correlation of biological activity and chemical structure, *J. Am. Chem. Soc.*, 86, 1616, 1964.
2. **Baughman, G. L. and Paris, D. F.,** Microbial bioconcentration of organic pollutants from aquatic systems — a critical review, *CRC Crit. Rev. Microbiol.*, Jan., 205, 1981.
3. **Sondergren, A.,** Uptake and accumulation of C^{14}-DDT by *Chlorella* sp. (Chlorophyceae), *Oikos*, 19, 126, 1968.
4. **Kerr, S. R. and Vass, W. P.,** Pesticide residues in aquatic invertebrates, in Edwards, C. A., Ed., *Environmental Pollution by Pesticides*, Plenum Press, London, 1973, 134.
5. **Crosby, B. G. and Tucker, R. K.,** Accumulation DDT by *Daphnia magna*, *Environ. Sci. Technol.*, 5, 714, 1971.
6. **Harding, G. C. H. and Vass, W. P.,** Uptake from sea water and clearance of DDT by marine planktonic crustacea, *J. Fish. Res. Board Can.*, 36, 247, 1979.
7. **Harding, G. C. and Vass, W. P.,** Uptake from sea water and clearance of DDT by the marine copepod *Calanus finmarchicus*, *J. Fish. Res. Board Can.*, 34, 177, 1977.
8. **Johnson, B. T., Saunders, C. R., Sanders, H. O., and Campbell, R. S.,** Biological magnification and degradation of DDT and aldrin by fresh water invertebrates, *J. Fish. Res. Board Can.*, 28, 705, 1971.
9. **Wilkes, F. G. and Weiss, C. M.,** The accumulation of DDT by the dragonfly nymph, *Tetragoneuria*, *Trans. Am. Fish. Soc.*, 100, 222, 1971.

10. **Pasteels, J. J.**, Pinocytose et athrocytose par 1-epithelium brachial de *Mytilus edulis*, *Z. Zellforsch.*, 92, 239, 1968.

11. **Geyer, H., Sheehan, D., Kotzias, D., Freitag, D., and Korte, F.**, Prediction of ecotoxicological behaviour of chemicals: relationship between physicochemical properties and bioaccumulation of organic compounds in the mussel, *Chemosphere*, 11, 1121, 1982.

12. **Ernst, W.**, Determination of the bioconcentration potential of marine organisms — a steady state approach. I. Bioconcentration data for seven chlorinated pesticides in mussels (*Mytilus edulis*) and their relation to solubility data, *Chemosphere*, 11, 731, 1977.

13. **Oliver, B. G.**, Biouptake of chlorinated hydrocarbons from laboratory-spiked and field sediments by oligochaete worms, *Environ. Sci. Technol.*, 21, 785, 1987.

14. **Shaw, G. R. and Connell, D. W.**, Comparative kinetics for bioaccumulation of polychlorinated biphenyls by the polychaete *(Capitella capitata)* and fish *(Mugil cephalus)*, *Ecotoxicol. Environ. Saf.*, 13, 84, 1987.

15. **Ferguson, D. E., Ludke, J. L., and Murphy, G. G.**, Dynamics of endrin uptake and release by resistant and susceptible strains of mosquito fish, *Trans. Am. Fish. Soc.*, 95, 335, 1966.

16. **Gakstatter, J. A. and Weiss, C. M.**, The elimination of DDT, dieldrin and lindane from fish following a single sublethal exposure in aquaria, *Trans. Am. Fish. Soc.*, 96, 301, 1967.

17. **Chadwick, G. G. and Brocksen, R. W.**, Accumulation of dieldrin by fish and selected fish-food organisms, *J. Wildl. Manage.*, 33, 693, 1969.

18. **Hamelink, J. L., Waybrant, R. C., and Ball, R. C.**, A proposal: exchange equilibria control the degree chlorinated hydrocarbons biologically magnify in lentic environments, *Trans. Am. Fish. Soc.*, 100, 207, 1971.

19. **Reinert, R. E.**, Accumulation of dieldrin in alga, *Daphnia magna* and the guppy, *J. Fish. Res. Board Can.*, 29, 1413, 1972.

20. **Connell, D. W.**, Bioaccumulation behaviour of persistent organic chemicals with aquatic organisms, *Rev. Environ. Contam. Toxicol.*, 101, 117, 1988.

21. **Holden, A. V.**, A study of the absorption of C-labelled DDT from water by fish, *Ann. Appl. Biol.*, 50, 467, 1962.

22. **Murphy, P. G. and Murphy, J. V.**, Correlations between respiration and direct uptake of DDT in the mosquito fish, *Bull. Environ. Contam. Toxicol.*, 6, 581, 1971.

23. **Kikuchi, M., Wakabayashi, M., Kojima, H., and Yoshida, T.**, Uptake, distribution and elimination of sodium linear alkylbenzene sulfonate and sodium alkyl sulphate in carp, *Ecotoxicol. Environ. Saf.*, 2, 115, 1978.

24. **Grzenda, A. R., Paris, D. F., and Taylor, W. J.**, The uptake, metabolism and elimination of chlorinated residues by goldfish *(Carassius auratus)* fed a DDT contaminated diet, *Trans. Am. Fish. Soc.*, 99, 385, 1970.

25. **Connell, D. W.**, A kerosene-like taint in the sea mullet *Mugil cephalus* (linneaus) II. Some aspects of the deposition and metabolism of hydrocarbons in muscle tissue, *Bull. Environ. Contam. Toxicol.*, 20, 492, 1978.

26. **Muir, D. C. G., Yarrechewski, A. L., and Knoll, A.**, Bioconcentration and deposition of 1,3,6,8-tetrachlorodibenzo-p-dioxin and octachlorodibenzo-p-dioxin by rainbow trout and flathead minnows, *Environ. Toxicol. Chem.*, 5, 261, 1986.

27. **Fong, W. C.**, Uptake and retention of Kuwait crude oil and its effect on oxygen uptake by the soft-shell clam *Mya arenaria*, *J. Fish. Res. Board Can.*, 33, 2774, 1976.

28. **Solbakken, J. E., Jeffery, F. M. H., Knapp, A. H., and Palmork, K. H.**, Accumulation and elimination of phenanthrene in the calico clam *(Macrocollista maculata)*, *Bull. Environ. Contam. Toxicol.*, 28, 530, 1982.

29. **Boryslawskyj, M., Garrood, A. C., and Pearson, J. T.**, Rates of accumulation of dieldrin by a freshwater filter feeder- *Sphaerium corneum*, *Environ. Pollut.*, 43, 3, 1978.

30. **Goerke, H.**, Temperature dependent elimination of 2,4,6,2',4'-pentachlorobiphenyl in *Nereis virens* (polychaeta), *Arch. Environ. Contam. Toxicol.*, 13, 347, 1984.

31. **Geyer, H., Scheunert, T., and Korte, F.**, Relationship between the lipid content of fish and their bioconcentration potential of 1,2,4-trichlorobenzene, *Chemosphere*, 14, 545, 1985.

32. **Zitko, V.**, Metabolism and distribution by aquatic animals, in *Handbook of Environmental Chemistry*, Hutinger, O., Ed., Springer-Verlag, Berlin, 221, 1980.

33. **Connell, D. W. and Hawker, D. W.**, Use of polynomial expressions to describe the bioconcentration of hydrophobic chemicals by fish, *Ecotoxicol. Environ. Saf.*, 16, 242, 1988.

34. **Neely, W. B., Branson, D. R., and Blau, G. E.**, Partition coefficients to measure bioconcentration potential of organic chemicals in fish, *Environ. Sci. Technol.*, 8, 1113, 1974.

35. **Kenaga, E. E. and Goring, C. A.**, Relationship between water solubility, soil sorption, octanol water partitioning and bioconcentration of chemicals in biota, in *Aquatic Toxicology*, Eaton, J. G., Parrish, P. R., and Hendricks, A. C., Eds., Vol. 707, American Society for Testing and Materials, Philadelphia, 78, 1980.

36. **Lu, P. Y. and Metcalf, R. L.**, Environmental fate and biodegradation biodegradability of benzene derivatives as studied in a model aquatic ecosystem, *Environ. Health Perspect.*, 10, 269, 1975.
37. **Ernst, W.**, Determination of the bioconcentration potential of marine organisms — a steady state approach. I. Bioconcentration data for seven chlorinated pesticides in mussels and their relation to solubility data, *Chemosphere*, 11, 731, 1977.
38. **Mackay, D.**, Correlation of bioconcentration factors, *Environ. Sci. Technol.*, 16, 274, 1982.
39. **Chiou, C. P.**, Partition coefficients of organic compounds in lipid-water systems and correlation with fish bioconcentration factors, *Environ. Sci. Technol.*, 19, 57, 1985.
40. **Dobbs, A. J. and Williams, N.**, Fat solubility — a property of environmental relevance? *Chemosphere*, 12, 97, 1983.
41. **Hansch, C.**, A quantitative approach to biochemical structure-activity relationships, *Acc. Chem. Res.*, 2, 232, 1969.
42. **Sugiura, K., Ito, N., Matsumoto, N., Mihara, Y., Murata, K., Tsukakoshi, Y., and Goto, M.**, Accumulation of polychlorinated biphenyls and polybrominated biphenyls in fish: limitation of "correlation between partition coefficients and accumulation factors", *Chemosphere*, 9, 731, 1978.
43. **Konemann, H. and van Leeuwen, H.**, Toxicokinetics in fish: accumulation and elimination of six chlorobenzenes by guppies, *Chemosphere*, 9, 3, 1980.
44. **Muir, D. C. G., Marshall, W. K., and Webster, G. R. B.**, Bioconcentration of PCD's by fish: effects of molecular structure and water chemistry, *Chemosphere*, 14, 829, 1985.
45. **Opperhuizen, A., Wagennar, W. J., Van der Wheilen, F. W. M., Van der Berg, Olie, K., and Gobas, F. A. P. C.**, Uptake and elimination of PCDD/PCDF congeners by fish after aqueous exposure to a fly-ash extract from municipal incinerator, *Chemosphere*, 15, 209, 1986.
46. **Bruggeman, W. A., Opperhuizen, A., Wijbenga, A., and Hutzinger, O.**, Bioaccumulation of super-lipophilic chemicals in fish, *Toxicol. Environ. Chem.*, 7, 176, 1984.
47. **Anliker, R. and Moser, P.**, The limits of bioaccumulation of organic pigments in fish: their relation to the partition coefficient and the solubility in water and octanol, *Ecotoxicol. Environ. Saf.*, 13, 43, 1987.
48. **Brooke, B. N., Dobbs, A. J., and Williams, N.**, Octanol: water partition coefficients (P): measurement, estimation, and interpretation, particularly for chemicals with $P \leqslant 10^5$, *Ecotoxicol. Environ. Saf.*, 11, 251, 1986.
49. **Miller, M. M., Wasik, S. P., Huang, G. L., Suiu, W. Y., and Mackay, D.**, Relationships between octanol-water partition coefficient and aqueous solubility, *Environ. Sci. Technol.*, 19, 522, 1985.
50. **Courtney, W. A. M. and Langston, W. J.**, Uptake of polychlorinated biphenyl (Aroclor 1254) from sediment and from seawater in two intertidal polychaetes, *Environ. Pollut.*, 15, 303, 1978.
51. **Oliver, B. G.**, Uptake of chlorinated organics from anthropogenically contaminated sediments by oligochaete worms, *Can. J. Fish. Aquat. Sci.*, 41, 878, 1984.
52. **Oliver, B. G.**, Biouptake of chlorinated hydrocarbons from laboratory-spiked and field sediments by oligochaete worms, *Environ. Sci. Technol.*, 21, 785, 1987.
53. **Ellgehausen, H., Guth, J. A., and Esser, H. O.**, Factors determining the bioaccumulation potential of pesticides in the individual compartments of aquatic food chains, *Ecotoxicol. Environ. Saf.*, 4, 134, 1980.
54. **Geyer, H., Politzki, G., and Freitag, D.**, Prediction of ecotoxicological behaviour of chemicals: relationship between n-octanol/water partition coefficient and bioaccumulation of organic chemicals by alga *Chlorella*, *Chemosphere*, 13, 269, 1984.
55. **Steen, W. C. and Karickhoff, S. W.**, Biosorption of hydrophobic organic pollutants by mixed microbial populations, *Chemosphere*, 10, 27, 1981.
56. **Casserly, D. M., Davis, E. M., Downs, T. D., and Guthrie, R. K.**, Sorption of organics by *Selenastrium Capricornutum*, *Water Res.*, 17, 1591, 1983.
57. **Mailhot, H.**, Prediction of algal bioaccumulation and uptake rate of nine organic compounds by ten physicochemical properties, *Environ. Sci. Technol.*, 21, 1009, 1987.
58. **Geyer, H., Viswinathan, R., Freitag, D., and Korte, F.**, Relationship between water solubility of organic chemicals and their bioaccumulation by the alga *Chlorella*, *Chemosphere*, 10, 1307, 1981.
59. **Southworth, G. R., Beachamp, J. J., and Schmider, P. K.**, Bioaccumulation potential of polycyclic aromatic hydrocarbons in *Daphnia Pulex*, *Water Res.*, 12, 973, 1978.
60. **Govers, H., Ruepert, C., and Aiking, H.**, Quantitative structure-activity relationships for polycyclic aromatic hydrocarbons: relation between molecular connectivity, physicochemical properties, bioconcentration and toxicity, in *Daphna pulex*, *Chemosphere*, 13, 227, 1984.
61. **Hawker, D. W. and Connell, D. W.**, Bioconcentration of lipophilic compounds by some aquatic organisms, *Ecotoxicol. Environ. Saf.*, 11, 184, 1986.
62. **Zhang, Y., Rott, B., and Freitag, D.**, Accumulation and elimination of PCB's by *Daphna magna* Strauss 1820, *Chemosphere*, 12, 1645, 1983.
63. **Gossett, R. W., Brown, D. A., and Young, D. R.**, Predicting the bioaccumulation of organic compounds in marine organisms using octanol/water partition coefficients, *Mar. Pollut. Bull.*, 14, 387, 1983.

64. **Clark, J. B., Patrick, J. M., Moore, J. C., and Forester, J.,** Accumulation of sediment-bound PCBs by fiddler crabs, *Environ. Contam. Toxicol.,* 36, 571, 1986.

65. **Connell, D. W., Bowman, M., and Hawker, D. W.,** Bioconcentration of chlorinated hydrocarbons from sediment by oligochaetes, *Ecotoxicol. Environ. Saf.,* 16, 293, 1988.

66. **Metcalf, R. L., Sanbourn, J. R., Lu, P. Y., and Nye, B.,** Laboratory model ecosystem studies of the degradation and fate of radiolabeled tri-tetra-, and pentachlorobiphenyl compared with DDE, *Arch. Environ. Contam.,* 3, 151, 1975.

67. **Veith, G. D., De Foe, D. L., and Bengstedt, B. V.,** Measuring and estimating the bioconcentration factor in fish, *J. Fish. Res. Board Can.,* 36, 1040, 1979.

68. **Veith, G. D., Macek, K. J., Petrocelli, S. R., and Carrol, J.,** An evaluation of using partition coefficients and water solubility to estimate bioconcentration factors for organic chemicals in fish, *Aquatic Toxicology,* ASTM STP707, Eaton, J. G., Parrish, P. R., Hendricks, A. C., Eds., American Society for Testing and Materials, 1980, 116.

69. **Ellgehausen, H., Guth, J. A., and Esser, H. O.,** Factors determining the bioaccumulation potential of pesticides in the individual compartments of aquatic food chains, *Ecotoxicol. Environ. Saf.,* 4, 134, 1980.

70. **Konemann, H. and van Leeuwen, K.,** Toxicokinetics of fish: accumulation and elimination of six chlorobenzenes by guppies, *Chemosphere,* 9, 3, 1980.

71. **Davies, R. P. and Dobbs, A. J.,** The prediction of bioconcentration in fish, *Water Res.,* 18, 1253, 1984.

72. **Saarikowski, J. and Viluksela, M.,** Relation between physicochemical properties of phenols and their toxicity and accumulation in fish, *Ecotoxicol. Environ. Saf.,* 6, 501, 1982.

73. **Oliver, B. G. and Niimi, A. J.,** Bioconcentration of chlorobenzenes from water by rainbow trout: correlations with partition coefficients and environmental residues, *Environ. Sci. Technol.,* 17, 287, 1983.

74. **Veith, G. G. and Kosian, P.,** Estimating bioconcentration potential from octanol/water partition coefficients, in *Physical Behaviour of PCBs in the Great Lakes,* Mackay, D., Paterson, S., and Eisenreich, S. J., Eds., Ann Arbor Science, Ann Arbor, MI, 1983, 269.

75. **Oliver, B. G.,** The relationship between bioconcentration factor in rainbow trout and physical-chemical properties for some halogenated compounds, in *QSAR in Environmental Toxicology,* Kaiser, K. I. E., Ed., D. Reidel Publishing, Dordrecht, Netherlands, 1984, 301.

76. **Oliver, B. G. and Niimi, A. J.,** Bioconcentration factors of some halogenated organics for rainbow trout: limitations in their use for prediction of environmental residues, *Environ. Sci. Technol.,* 19, 842, 1985.

77. **Zaroogian, G. E., Heltsche, J. F., and Johnson, M.,** Estimation of bioconcentration in marine species using structure-activity models, *Environ. Toxicol. Chem.,* 4, 3, 1985.

78. **Connell, D. W. and Schüürmann, G.,** Evaluation of various molecular parameters as predictors of bioconcentration in fish, *Ecotoxicol. Environ. Saf.,* 15, 324, 1988.

79. **Schüürmann, G. and Klein, W.,** Advances in bioconcentration prediction, *Chemosphere,* 17, 1551, 1988.

80. **Metcalf, R. L., Kapoor, I. P., Lu, P. Y., Schuth, C. K., and Scherman, P.,** Model ecosystem studies of the environmental fate of 6 organochlorine pesticides, *Environ. Health Perspect.,* 35, 44, 1973.

81. **Chiou, C. T., Freed, V. H., Schmedding, D. W., and Kohnert, R. L.,** Partition coefficient and bioaccumulation of selected organic chemicals, *Environ. Sci. Technol.,* 11, 475, 1977.

82. **Lu, P. Y. and Metcalf, R. L.,** Environmental fate and biodegradability of benzene derivatives as studied in a model aquatic ecosystem, *Environ. Health Perspect.,* 10, 269, 1975.

83. **Connell, D. W., Gabric, A., and Markwell, R.,** unpublished.

84. **Kenaga, E. E.,** Correlation of bioconcentration factors of chemicals in aquatic and terrestrial organisms with their physical and chemical properties, *Environ. Sci. Technol.,* 14, 553, 1980.

85. **Briggs, G. G.,** Theoretical and experimental relationships between soil adsorption, octanol-water partition coefficients, water solubilities, bioconcentration factors and the parachor, *J. Agric. Food Chem.,* 29, 1050, 1981.

86. **Koch, R.,** Molecular connectivity index for assessing ecotoxicological behaviour of organic compounds, *Toxicol. Environ. Chem.,* 6, 87, 1983.

87. **Muir, B. C. G., Yarechewski, A. L., and Grift, N. P.,** Environmental dynamics of phosphate esters. III. Comparison of the bioconcentration of four triaryl phosphates by fish, *Chemosphere,* 12, 155, 1983.

88. **Brooke, D. M., Dobbs, A. J., and Williams, N.,** Octanol: water partition coefficients P: measurement, estimation and interpretation, particularly for chemicals with $P < 10^5$, *Ecotoxicol. Environ. Saf.,* 11, 251, 1986.

89. **Sabljic, A. and Protic, M.,** Molecular connectivity: a novel method for prediction of bioconcentration factor of hazardous chemicals, *Chem. Biol. Interactions,* 42, 301, 1982.

90. **Ogata, M., Fujisawa, K., Ogino, Y., and Mano, E.,** Partition coefficients as measure of bioconcentration potential of crude oil compounds in fish and shell fish, *Bull. Environ. Contam. Toxicol.,* 33, 561, 1984.

91. **Karickhoff, S. W.,** Pollutant sorption in environmental systems, in *Environmental Exposure from Chemicals,* Neely, W. B. and Blau, G. E., Eds., CRC Press, Boca Raton, FL, 1985, 49.

92. **Chiou, C. T., Peters, L. J., and Freed, V. H.,** A physical concept of soil-water equilibria for non-ionic organic compounds, *Science,* 206, 831, 1979.

93. **Sabljic, A.,** Predictions of the nature and strength of soil sorption of organic pollutants by molecular topology, *J. Agric. Food Chem.*, 32, 243, 1984.

94. **Farrington, J. W. and Westall, J.,** Organic chemical pollutants in the oceans and groundwater: A review of fundamental chemical properties and biogeochemistry, in *The Role of the Oceans as a Waste Disposal Option*, Kullenberg, G., Ed., D. Reidel Publishing, Dordrecht, Netherlands, 1986, 361.

95. **Atkins, G. L.,** *Multi-compartment Models for Biological Systems*, Methuen, London, 1969.

96. **Moriarty, F.,** Exposure and residues, in *Organochlorine Insecticides: Persistent Organic Pollutants*, Moriarty, F., Ed., Academic Press, London, 1975, 29.

97. **Walker, C. H.,** Kinetic models for predicting bioaccumulation of pollutants in ecosystems, *Environ. Pollut.*, 44, 227, 1987.

98. **Spacie, A. and Hamelink, J. L.,** Alternative models for describing the bioconcentration of organics in fish, *Environ. Toxicol. Chem.*, 1, 309, 1982.

99. **Neely, W. B.,** Estimating rate constants for the uptake and clearance of chemicals by fish, *Environ. Sci. Technol.*, 13, 1606, 1979.

100. **Murphy, P. G. and Murphy, J. V.,** Correlations between respiration and direct uptake of DDT in the mosquito fish, *Bull. Environ. Contam. Toxicol.*, 6, 581, 1971.

101. **Konemann, H. and van Leeuwen, K.,** Toxicokinetics in fish: accumulation and elimination of six chlorobenzenes by guppies, *Chemosphere*, 9, 3, 1980.

102. **Hawker, D. W. and Connell, D. W.,** Relationships between partition coefficient, uptake rate constant, clearance rate constant and time to equilibrium for bioaccumulation, *Chemosphere*, 14, 1205, 1985.

103. **Gobas, F. A. P. C., Opperhuizen, A., and Hutzinger, O.,** Bioconcentration of hydrophobic chemicals in fish: relation with membrane permeation, *Environ. Toxicol. Chem.*, 5, 637, 1986.

104. **Hawker, D. W. and Connell, D. W.,** Influence of partition coefficient of lipophilic compounds on bioconcentration kinetics with fish, *Water Res.*, 22, 701, 1988.

105. **Shaw, G. R. and Connell, D. W.,** Physicochemical properties controlling polychlorinated biphenyl concentrations in aquatic organisms, *Environ. Sci. Technol.*, 18, 18, 1984.

106. **Opperhuizen, A., Velde, E. W., Gobas, F. A. P. C., Llem, D. A. K., and Steen, J. M. D.,** Relationship between bioconcentration in fish and steric factors of hydrophobic chemicals, *Chemosphere*, 14, 1871, 1985.

107. **Spigarelli, S. A., Thomas, M. M., and Prepejchal, W.,** Thermal and metabolic factors affecting PCB uptake by adult brown trout, *Environ. Sci. Technol.*, 17, 88, 1983.

108. **Connell, D. W. and Miller, G. J.,** *Chemistry and Ecotoxicology of Pollution*, John Wiley & Sons, New York, 1984.

109. **Esser, H. O. and Moser, P.,** An appraisal of problems related to the measurement and valuation of bioaccumulation, *Ecotoxicol. Environ. Saf.*, 6, 13, 1982.

110. **Murphy, P. G.,** Effects of salinity on uptake of DDT, DDE and DDD by fish, *Bull. Environ. Contam. Toxicol.*, 5, 404, 1970.

111. **Shaw, G. R. and Connell, D. W.,** Factors influencing concentrations of polychlorinated biphenyls in organisms from an estuarine ecosystem, *Aust. J. Mar. Freshwater Res.*, 333, 1057, 1982.

112. **Landrum, P. F., Reinhold, M. D., Nihart, F. R., and Eadie, B. J.,** Predicting the bioavailability of organic xenobiotics to *Pontoporeia hoyi* in the presence of humic and fulvic materials and natural dissolved organic matter, *Environ. Toxicol. Chem.*, 4, 459, 1985.

113. **Connell, D. W.,** Age to PCB concentration relationship with the striped bass *(Morone Saxatilis)* in the Hudson River and Long Island Sound, *Chemosphere*, 16, 1469, 1987.

114. **Sondergren, A.,** Solvent-filled dialysis membranes simulate uptake of pollutants by aquatic organisms, *Environ. Sci. Technol.*, 21, 855, 1987.

Chapter 7

BIOMAGNIFICATION OF LIPOPHILIC COMPOUNDS IN TERRESTRIAL AND AQUATIC SYSTEMS

Des W. Connell

TABLE OF CONTENTS

I. INTRODUCTION

The pathways of distribution of a compound and biotic uptake routes were discussed in Chapter 4 and are briefly outlined in broad terms in Chapter 4, Figure 1. Bioconcentration is one of the most important bioaccumulation processes and involves equilibrium between water and various biota, as was described in Chapter 6. A number of important bioaccumulation pathways are not bioconcentration processes; these are illustrated diagrammatically in Figure 1. Bioconcentration is also included in this illustration by the transfers to biota from water and interstitial water.

In this publication, biomagnification is defined as transfer of a chemical from food to a consumer, and was outlined in broad terms in Chapter 4. Table 1 illustrates examples of biomagnification in both terrestrial and aquatic systems; these are included in Figure 1 as the processes designated with single headed arrows. Biomagnification could occur as a result of a single food-to-consumer step, or as a result of a sequence of such steps. For example, transfer from lower trophic aquatic biota to higher trophic level aquatic biota, to higher level terrestrial biota, may occur, and does not necessarily involve an increase in concentration of the xenobiotic chemical at each step.

All of these biomagnification processes cannot be simply characterized by partition behavior, as with bioconcentration. The atmosphere could be suggested as a comparable major phase to water with aquatic organisms, and thus imply that it is the major influence on terrestrial biota. But this does not appear to be the situation, although the atmosphere is an important phase under some circumstances. The composition and nature of the food and feeding behavior seem to be major factors influencing biomagnification. Partition processes are involved, but these are more complex than those which occur with bioconcentration, and are not as easily defined and characterized.

In accord with food as the source of xenobiotic chemical and Biomagnification Factor (BF) is defined as the ratio of the concentration in the consumer (C_c) to the concentration in food (C_F).

$$BF = C_c/C_F$$

Some aspects of the nature of this factor and the techniques used to measure it have been outlined in Chapter 2. Recently, Travis and Arms[1] have suggested that a Biotransfer Factor (B) is a more useful evaluation of the biomagnification property of chemicals. This is defined as the ratio between the concentration in the animal and the daily intake of the chemical.

Examination of Figure 1 indicates that some chemical transfer processes, in particular those between biota and the atmosphere, can be characterized by partition behavior. These are described by doubleheaded arrows, as illustrated in Figure 1 with examples in Table 2. This partitioning behavior is suggested by the relatively free movement of chemical molecules from the atmosphere to biota and in the reverse direction. These transfers are thus not analogous to biomagnification, and would be expected to be characterized by different mechanisms and factors. In fact, the mechanism of these processes has more similarities to bioconcentration, which is an aquatic phenomenon. In Chapter 4 it is suggested that an appropriate terminology for these transfers is "terrestrial bioconcentration".

II. THE SOIL-TO-ATMOSPHERE SYSTEM

Many xenobiotic chemicals are discharged to the atmosphere, either deliberately or unintentionally. For example, pesticides are often broadcast throughout agricultural areas, and polyaromatic hydrocarbons are discharged in motor vehicle exhaust to the atmosphere. Much of the residues produced by these discharges are accumulated in the soil. For example,

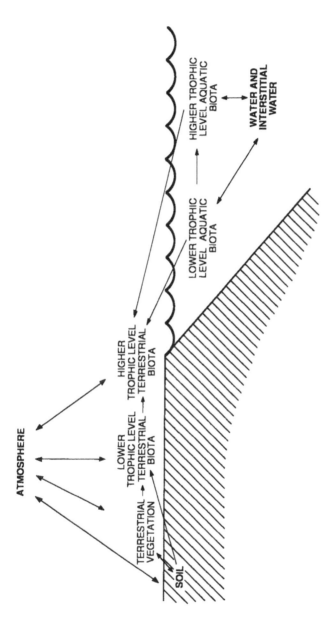

FIGURE 1. Transfer pathways of a persistent xenobiotic chemical in the terrestrial and aquatic environments. The double headed arrows indicate processes where there is direct equilibrium between the two components involved, whereas this does not occur in transfers indicated by a single headed arrow.

TABLE 1

Examples of Food-to-Consumer Relationships Which May Result in Biomagnification[a]

Food class	Examples	Consumer class	Examples
Low trophic level terrestrial biota	Earth worms, insects	Higher trophic level terrestrial biota	Terrestrial birds
Terrestrial vegetation	Grass, grains, trees	Low trophic level terrestrial biota	Mammals, insects, terrestrial birds
Aquatic infauna	Aquatic worms, other invertebrates	Higher tropic level terrestrial biota	Aquatic birds
Low trophic level aquatic biota	Mollucs, phytoplankton	Higher trophic level aquatic biota	Fish
Low and higher level aquatic biota	Fish, molluscs and other aquatic invertebrates	High trophic level aquatic biota	Aquatic birds, mammals

[a] Soil may be consumed accidentally or intentionally in food to give a soil-to-biota transfer with organisms such as terrestrial worms and grazing mammals.

TABLE 2

Examples of Pairs of Terrestrial Phases Where Transfer of Chemicals may Transfer from an Abiotic Component to a Biotic Component[a]

Abiotic component	Biotic component	Examples
Soil	Terrestrial vegetation	Trees, grass
Air	Terrestrial vegetation	Trees, grass
Air	Low and higher trophic level terrestrial biota	Mammals

[a] Described as "terrestrial bioconcentration".

in 1970, Korschgen[2] carried out a thorough investigation of the contamination of soils in an agricultural area by aldrin, and found an average concentration of 0.31 ppm. More recently, Butler et al.[3] have reported concentrations of polycyclic aromatic hydrocarbons (PAHs) in urban soils up to 20 ppm; also, Neilsen and Looke[4] have found significant concentrations of polychlorodibenzodioxins (PCDD) and polychlorodibenzofurans (PCDF) in various urban and rural soils.

The accumulation of a chemical in vapor form in the atmosphere onto soil in the soil-to-atmosphere system is an equilibrium process and can be described by a Freundlich type of equation. At low concentrations where the nonlinearity constant is unity, then

$$K_{SA} = C_S/C_A \tag{1}$$

where K_{SA} is the soil-to-atmosphere distribution coefficient, C_S is the concentration in soil, and C_A is the concentration in atmosphere. The C_A is often expressed as vapor density,[5] and using this measure, Freundlich-type relationships have been developed for a variety of chlorohydrocarbon insecticides, as shown by Spencer and Cliath[6] in Figure 2.

The mechanisms of sorption of chemicals on soil, in the soil-to-atmosphere system, would be expected to have some similarity to the soil-to-water system, but be distinctively different. If the sorption characteristics on soil on both systems were the same, then from Equation 1, and because $H = C_A/C_W$ and $K_D = C_S/C_W$, the following relationship can be obtained:

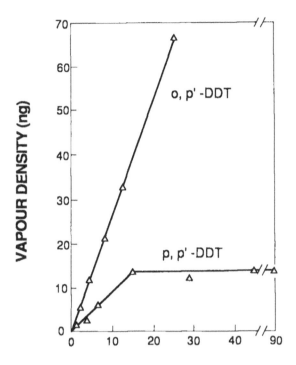

FIGURE 2. A Freundlich-type plot of the concentration of DDT in the atmosphere above a silt loam with 7.5% water (C_A, represented by Vapor Density) against concentration in soil (C_S). (From Spencer, W. F. and Cliath, M. M., *J. Agric. Food Chem.*, 20, 645, 1972. Copyright American Chemical Society. With permission.)

$$K_{SA} = K_D/H$$

where K_D is the soil-to-water partition coefficient, and H the Henry's Law Constant.

In fact, the presence of water in soil exerts a major influence on K_{SA} with lipophilic compounds. This is demonstrated with the relationship in Figure 3, which was reported by Spencer et al.[7] This data indicates that using a constant soil dieldrin concentration with low soil water concentrations, the vapor density of dieldrin is low due to the relatively strong sorption in the soil. However, as the soil water content increases, the concentration of pesticide in the atmosphere above the soil also increases, the concentration of pesticide in the atmosphere above the soil also increases, which would lead to a decline in K_{SA}. It is believed that the increase in concentration of water in the soil makes sorption of the compound increasingly difficult until the soil surfaces are saturated with water. At this point, about 0.028 g g^{-1} soil water content (see Figure 3), further additions of water have no affect and the atmospheric concentration of the compound is that exerted by the compound without the presence of soil at all.

In accord with these findings, Chiou and Shoup[8] have reported that atmospheric humidity has an influence on the soil sorption capacity of lipophilic compounds. These compounds have much higher soil sorption capacities at subsaturation humidities.

The organic carbon or organic matter content of the soil is an important factor in the soil-to-atmosphere system, but is modified by the soil moisture content. An illustrative set

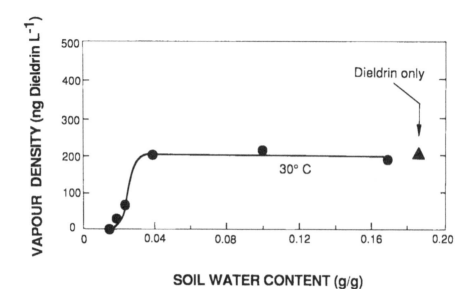

FIGURE 3.   Effect of soil water content on the concentration of dieldrin (C_A, represented by Vapor Density) in the atmosphere over silt loam soil containing 100 mg kg^{-1}. (From Spencer, W. F., Farmer, W. J., and Cliath, M. M., *Residue Rev.*, 49, 1, 1973. Copyright Springer-Verlag. With permission.)

TABLE 3
Effect of Organic Matter and Clay Content on Vapor Density of Dieldrin at 30°C in Wet and Dry Soils Containing Ten ppm of Dieldrin

Soil type	Organic matter (%)	Clay (%)	Vapor density (ng L^{-1})	
			Wet	Dry
Rosita very fine sandy loam	0.19	16.3	175	1.7
Imperical clay	0.20	67.3	200	2.9
Gila silt loam	0.58	18.4	52	0.7
Kentwood sandy loam	1.62	10.0	32	0.4
Linne clay loam	2.41	33.4	32	0.6

From Connell, D. W., *Rev. Environ. Contam. Toxicol.*, 101, 128, 1988. Copyright Springer-Verlag. With permission.)

of data is shown in Table 3. The dry soil exhibits a general trend of declining vapor density, and thus increasing K_{SA}, with increasing organic matter content of the soil. The percentage of clay content does not seem to have a significant influence on the soil-to-atmosphere partitioning, as would be expected with these lipophilic compounds.

Recently Chiou et al.[9] have shown that highly linear isotherms are obtained for the atmosphere-to-soil humic acid partition, and a parallel is drawn to the water-to-sediment organic matter partition. A comparison of the shape of the isotherms obtained in both systems suggests that water absorbed onto the exterior surfaces of the humic organic matter effectively suppresses the sorption of organic compounds.

It would be expected that contact and this ability to exchange compounds with the atmosphere would decline with vertical depth in the soil. This would be reflected in a declining concentration of lipophilic compounds where exposure to atmosphere xenobiotic compounds occurs. This pattern has been observed with a wide variety of compounds in different soil types, as illustrated by the results of Butler et al.[3] and Nielsen and Lokke.[4]

Such factors as soil porosity and exposure to water have an influence on the nature of this vertical soil concentration profile.

In contrast to equilibrium partitioning between soil and the atmosphere, the volatilization of compounds from soil to the atmosphere has received considerable attention. The rate of volatilization of lipophilic compounds depends on a variety of factors, such as the vapor pressure of the compounds, soil water content, and the amount of soil organic matter present. These factors have been reviewed by Thibodeaux,[10] and a laboratory method for evaluation of volatilization rates has been described by Kilzer et al.[11]

III. MECHANISMS AND QSARS FOR BIOMAGNIFICATION AND TERRESTRIAL BIOCONCENTRATION

A. ATMOSPHERE-TO-BIOTA TRANSFERS (TERRESTRIAL BIOCONCENTRATION)

1. Atmosphere-to-Plant System

A variety of plants, including spruce *(Picea abies)*, mosses *(Hypnum cupressiforme* and *Rhaconitrium)*, and pines *(Pinus silvestris* and *Pinus pinea)* have been used as biomonitors for lipophilic xenobiotic compounds in the atmosphere.[12,13] In addition, some of these, as well as a variety of other plants, have been used to monitor atmospheric metals and other substances. Buckley[14] found that the different plants species have different accumulation capacities for polychlorobiphenyls (PCBS) in the atmosphere, but the differential accumulation factor between species remains relatively constant. On the other hand, Gaggi et al.[13] present data indicating the suitability of plants as biomonitors for atmospheric lipophilic pollutants, and suggests that many plant species are comparable and suitable.

Many of the better known lipophilic xenobiotic compounds have been monitored using this technique. For example, Bacci et al.[15] used mosses and lichens to evaluate the occurrence of HCB, hexachlorocyclohexane isomers (HCH), p, p-'DDT, DDE, and PCB congeners in the Antarctic atmosphere. The same set of compounds was investigated in a variety of vegetation types from many countries by Gaggi et al.[13] Buckley[14] used foliage from various plants in North America for atmospheric PCB investigations. Mosses were found useful by Thomas[12] to evaluate atmospheric PAHs and benzenehexachlorides. Reischl et al.[16] analyzed conifer needles to evaluate the atmospheric occurrence of PCDD and PCDF.

It is believed that in most of these situations the xenobiotic chemical occurs as a vapor in the atmosphere or is absorbed onto atmospheric particulates. It is taken up by the wax cuticle on the other surface of the foliage of most plants by passive sorption processes[17] from compounds in the vapor form in the atmosphere, or from particulates deposited on the plant surface. Thus, Thomas et al.[18] have suggested that the mass-to-surface area is an important factor in plant bioaccumulation. In addition, these authors investigated a number of different plant species and found that lipophilic compounds were most efficiently bioaccumulated in plants which were relatively high in lipid matter. Thomas[12] has found multiple linear regression equations of the following type for uptake of PAHs by mosses:

$$C_1 = 1.62 \, C_2 + 259.60 \, C_3 + 0.74 \, P$$

(for benzo(ghi)perylene, $R = 0.74$ for $n = 15$; $p = 0.002$)

where C_1 is the concentration in moss (ng g^{-1}), C_2 is the concentration in rainwater (ng L^{-1}), C_3 is the concentration in atmospheric particulate matter (ng m^{-3}), and P is the amount of precipitation (mm).

This indicates that the xenobiotic chemical concentration in atmospheric particulate matter is a major influence on moss concentration. Thomas[12] suggests that this reflects the filtration capacity of the mosses for effective retention of particulate matter. This result also suggests that concentration in rainwater and the amount of precipitation are of relatively minor importance.

Gaggi and Bacci[19] have measured the rates of uptake of a variety of chlorinated hydrocarbons from the background levels of these substances in the Italian atmosphere onto pine needles. Also, Gaggi et al.[13] have estimated bioaccumulation factors for these chemicals from laboratory experiments with controlled vapor concentrations and azaleas. However, these results are not cited in this reference. Using foliage affinities and the pine needle accumulation rates mentioned previously, Gaggi et al.[13] were able to calculate the levels of some chlorinated hydrocarbon vapors in the Italian atmosphere.

At the present time, bioaccumulation factors relating atmosphere and vegetation concentration for lipophilic chemicals at equilibrium are not available. Connell and Hawker[20] recognized this difficulty in using the fugacity modeling approach to estimate environmental distribution. To overcome this problem, these authors assumed that partitioning takes place exclusively between vegetation lipid and the atmosphere, and developed the following expressions:

$$K_{VA} = C_V/C_A$$

where K_{VA} is the plant-to-atmosphere partition coefficient, C_V is the concentration in vegetation, and C_A is the concentration in air. Thus

$$K_{VA} = (C_V/C_W)(C_W/C_A)$$

If the C_V/C_W partition is equivalent to the vegetation lipid-to-water partition and a good surrogate for this process is the octanol-to-water process, then

$$C_V/C_W = L_V K_{OW}$$

where L_V is the fraction of lipid in the vegetation.

Since C_A/C_W is H, the Henry's law constant, then

$$K_{VA} = (L_V K_{OW})/H$$

the utility of this expression for estimating K_{VA} values is yet to be evaluated.

In some situations, the xenobiotic chemical originating from the atmosphere may be in the form of an aerosol or larger droplets. This applies particularly with agricultural pesticides; these substances in solution may actually wet the outer surface of the plant. Penetration, or deposition in the surface cuticle, may result from this situation. Kenaga[21] has reviewed processes of this type. In these circumstances, the surface-to-mass ratio may be a major factor influencing the overall concentration in a plant. Table 4 shows calculations reported by Coulston and Korte[22] which indicate the potential influence of these factors on residues present in plants.

2. Atmosphere-to-Animal System

Xenobiotic chemicals in the atmosphere are principally present as vapors, aerosols, and larger droplets, and are absorbed onto particulates. Chemicals can enter respiratory organs or the outer body surface, although the respiratory organs are usually most important with mammals, and skin or outer body surface most important with insects.

Aspects of the atmosphere-to-insect route of bioaccumulation has been the subject of many investigations. Usually, the xenobiotic chemical of interest has been an insecticide, and it is often contained in solution with chemical agents which enhance its uptake by insects.

TABLE 4
Limits to the Residue Concentrations on Plants Calculated from
the Surface Area to Weight Ratio[a]

Categories of plants	Concentration (ppm)	
	Upper limit	Typical limit
Range grass (short)	240	125
Grass (long)	110	92
Leaves and leafy crops (vegetables and fruit)	125	35
Forage crops (alfalfa, clover)	58	33
Pods containing seeds (legumes)	12	3
Fruit (cherries, peaches, grapes, citrus)	7	1.5

[a] Based on 1 pound per acre dosage.

From Coulston, F. and Korte, F. *Environmental Quality and Safety*, Vol. I *Global Aspects of Chemistry, Toxicology and Technology as Applied to the Environment*, Georg Thieme Verlag, Stuttgart, 1972, 17. Copyright Georg Thieme Verlag, Stuttgart.

In addition, the chemical and other agents can be present in the atmosphere as aerosols and larger droplets. The outer surfaces of many insects are often actually wet with the insecticide solution, and bioaccumulation then occurs through deposition in, or penetration through, the insect cuticle. For example, Davis and French[23] found that ground beetles exposed to spraying in fields contained 70 ppm on their body surfaces, whereas those in the area not directly exposed contained a maximum of 4 ppm.

Similarly, with mammals bioaccumulation can occur by direct absorption of the chemical through the skin after wetting. A more common form for the uptake of many xenobiotic chemicals is as a vapor, or sorbed onto atmospheric particulates. Filov et al.[24] have briefly reviewed some aspects of these processes. Lipophilic gases and vapors pass through the alveoli of the lungs into the bloodstream, and then to all parts of the body. For example, Stacey and Tatum[25] have found evidence for uptake of organochlorine insecticides in the vapor form by human subjects, and the deposition of these substances in human fat, particularly milk. The uptake process is described by an approximately exponential expression, and the uptake approaches a steady state after an equilibration time period. The ratio of the concentration of compound in the inspired and expired air serves as a guide to the establishment of equilibrium in mammals, as shown in Figure 4. The process of biodegradation, in addition to uptake and distribution, may lead to an equilibrium ratio of less than unity.

Bioaccumulation factors, i.e., concentration in the animal/concentration in the atmosphere, for the atmosphere-to-organism bioaccumulation process have not been commonly measured. Most biological properties, e.g., toxicity, narcosis, and so on, have been related to the environmental concentration. However, these concentrations have been found to be commonly related to the carbon number of the compounds involved, suggesting a possible relationship between such factors as K_{AW}, water solubility, and the atmosphere-to-animal partition coefficient. In an idealized and simplified situation, an expression can be developed for the animal-to-atmosphere partition coefficient at equilibrium. The animal-to-atmosphere partition coefficient can be defined as

$$K_{AA} = C_{AN}/C_A$$

where K_{AA} is the animal-to-atmosphere partition coefficient at equilibrium, C_{AN} is the concentration in the animal, and C_A is the concentration in air. Thus

$$K_{AA} = (C_{AN}/C_W)(C_W/C_A)$$

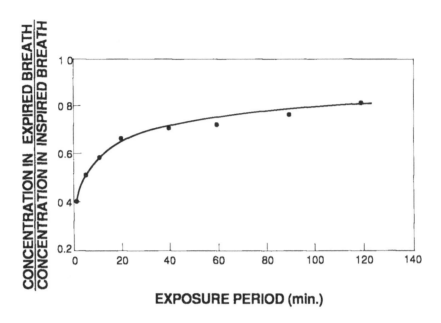

EXPOSURE PERIOD (min.)

FIGURE 4. Ratio of the concentrations of a halothane in the inspired and expired breath of human subjects. (Adapted from Filov, V. A., Golubev, A. A., Liublina, E. L. and Tolokontsev, *Quantitative Toxicology*, John Wiley & Sons, New York, 1979.)

If the C_{AN}/C_W factor is analogous to the bioconcentration of lipophilic compounds from water by aquatic organisms, then

$$C_{AN}/C_W = L_A K_{OW}$$

where L_A is the lipid content of the animal.

Since C_A/C_W is the Henry's law constant, then

$$K_{AA} = (L_A K_{OW})/H$$

Whether the animal-to-atmosphere partition coefficient can be obtained by this simple expression remains to be evaluated.

Xenobiotic chemicals sorbed onto atmospheric particulates may be deposited in parts of the respiratory system and bioaccumulated. The zone of deposition depends on such factors as size, shape, change, etc. Pott and Oberdorster[26] have reviewed the information available on these processes with regard to PAHs. After deposition of the particulate matter the sorbed PAHs can be translocated to various parts of the organism.

B. SOIL-TO-BIOTA TRANSFERS (TERRESTRIAL BIOCONCENTRATION)
1. Soil-to-Vegetation Processes

There are two possible major routes of transfer of environmental xenobiotic chemicals to plants. First, from the soil through the roots, and second, through sorption on the foliage from the atmosphere. In this chapter, Section III.A., the direct sorption of vapors from the atmosphere by vegetation was outlined. This section will be concerned with the soil-to-atmosphere to plant foliage process, as well as transfer through the roots.

Nash and Beall[27] conducted experiments on soybean plants and concluded that DDT residues in foliage arose from vapor movement from contaminated soil, but dieldrin, endrin and heptachlor residues arose from root uptake. With PCBs, Webber and Morzek[28] found

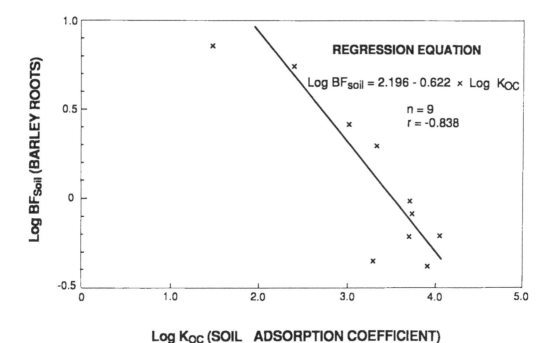

FIGURE 5. Relationship between barely root concentration factor (BF_{soil} is concentration in barley roots wet weight per concentration in soil air dried weight). The soil adsorption coefficient based on organic carbon content (K_{OC} is g adsorbed per kilogram soil solids per gram dissolved per liter of water × 1 per organic carbon fraction) for a variety of lipophilic compounds. (From Topp, E., Schevnent, A., Attar, A., and Korte, F., *Ecotoxicol. Environ. Saf.*, 11, 219, 198. Copyright Academic Press. With permission.)

that soybean translocated little to the foliage from soil, although fescue translocated more. In reviewing the range of data available, Strek and Weber[29] concluded that PCBs translocated from soil to plant foliage depending on the species involved, degree of chlorination PCBs, and growth period of the plant. However, the amount of translocation through the root system was generally low.

Iwata et al.[30] and several other authors have provided evidence that lipophilic compounds in soil are sorbed onto the outer surface of the roots of several plants, and translocation internally of these substances may be very low. Strek and Weber,[29] in collating the available data, suggested that carrot roots may have the greatest root capacity of the plants investigated to accumulate PCBs.

Using barley plants, Topp et al.[31] observed a correlation between log BF_{soil} of roots, i.e., concentration in roots, wet weight/concentration in soil, air-dried weight, and the soil sorption coefficient, K_{OC}, for a range of chemicals, including several chlorinated hydrocarbons.

$$\log BF_{soil} = 2.20 - 0.62 \log K_{OC} \qquad (n = 9; \quad r = -0.838)$$

This relationship is negative, as shown in Figure 5, because bioaccumulation occurs only from the relatively small proportion of compounds in soil water which is available for uptake. This process is somewhat analogous to uptake by aquatic infauna, as described in Chapter 6, Section II.A.3. and Section II.B.3. This suggests that the desorption from soil to water and the absorption from water to root tend to cancel one another out, and the concentrations in soil and root should be somewhat similar. This is apparent from the limited linear range of values which BF_{soil} covers, with equivalent concentrations in soil and roots occurring in a centralized position where log BF is zero (see Figure 5).

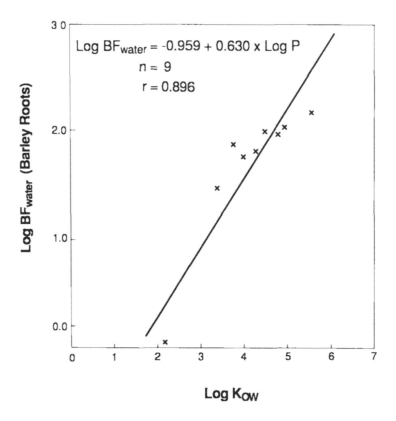

FIGURE 6.   Relationship between barely root concentration factor (BF_{water} is concentration in roots, wet weight per concentration in soil water) and the octanol-to-water partition coefficient (K_{OW}) for a variety of lipophilic compounds. (From Topp, E., Schievner, A., Attar, A., and Korte, F., *Ecotoxicol. Environ. Saf.*, 11, 219, 1986. Copyright Academic Press. With permission.)

Topp et al.[31] have calculated the soil water concentration, and from this BF values (concentration in roots, wet weight/concentration in soil water) were determined. The relationship in Figure 6 was obtained from this data. In this situation the bioaccumulation factor is expressed in terms of water concentration, and thus exhibits a comparatively wide range of values. The linear regression equation found was

$$\log BF_{water} = -0.96 + 0.63 \log K_{ow} \qquad (n = 9; \ r = 0.896)$$

The expected positive relationship between BF_{water} and K_{ow} is evident in this equation.

More recently, Bacci and Gaggi,[32,33] have conducted a variety of experiments on roots relating to the bioaccumulation of lipophilic compounds in various plants. They concluded that the compounds in the foliage originate from the soil-to-atmosphere-to-foliage transfer process. The amount of compound translocated from roots to foliage resulting from bioaccumulation from soil was relatively minor. Also, an investigation of barley foliar bioaccumulation from soil by Topp et al.[31] revealed a correlation between foliar uptake and volatilization from soil.

Topp et al.[31] decided to investigate the relationship of total bioaccumulation of lipophilic chemicals, i.e., both foliar and root bioaccumulation, to molecular weight. The relationship shown in Figure 7 was obtained with a linear regression equation of

$$\log BF = 5.94 - 2.39 \log M \qquad (n = 14; \ r = 0.9449)$$

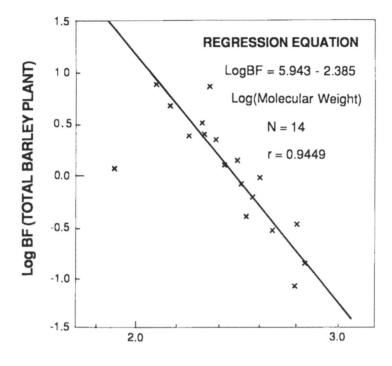

FIGURE 7. Relationship between total barley plant bioconcentration factor (BF is total concentration in plant, wet weight per concentration in soil, air-dried weight) and molecular weight for a variety of lipophilic compounds. (From Topp, E., Schevnert, A., Attar, A., and Korte, F., *Ecotoxicol. Environ. Saf.*, 11, 219, 1986. Copyright Academic Press. With permission.)

where BF is the concentration in the total plant, wet weight/concentration in soil, air dried weight, and M is the molecular weight.

The compounds were mainly chlorinated hydrocarbons, but some pigments were included which are considered highly nonvolatile. The authors concluded that molecular weight is a more suitable parameter to predict plant bioaccumulation than the octanol-to-water partition coefficient. However, they also point out that barley is relatively low in lipid, and some other plants are comparatively high, such as carrot, cress, and so on, and thus may exhibit higher bioaccumulation factors.

Travis and Arms[1] have investigated the bioaccumulation of a variety of organic chemicals, mainly chlorinated hydrocarbons, but including other chemical types of pesticide, in vegetation from soil. A somewhat similar linear regression equation to that above for molecular weight was obtained using K_{ow} values, thus

$$\log BF = 1.59 - 0.58 \log K_{ow} \qquad (n = 29, \quad r = 0.73)$$

They concluded that this inverse proportionality is not surprising, since transport from soil to the atmosphere is dependent on the solubility of a chemical in water which is inversely proportional K_{ow}.

Both of these relationships show little increase in concentration in foliage above that present in soil. In fact, most of the compounds considered by Travis and Arms[1] exhibit lower concentrations in foliage than are present in the soil. Similarly, results of Topp et

TABLE 5
Bioaccumulation Factors (BF) Observed with Soil Invertebrates[a]

Organism	Compound	BF	Ref.
Earthworms (*Allolobophora* sp.)	Aldrin and dieldrin	4.8	2
Crickets *(Gryllus assimilis)*	Aldrin and dieldrin	<1	2
Ground beetles *(Harpalus* sp.)	Aldrin and dieldrin	3.5	2
Ground beetles *(Poecilus* sp.)	Aldrin and dieldrin	31	2
Earthworms	TCDD	0.4—3.5	34
Earthworms	Tetrachloro- and hexachloro-benzene	0.12—19.5	34
Earthworms	Various insecticides	<1—10	35
Slug (*Peroceras reticulatum*)	HCB	0.4—1.4	36

[a]   BF, concentration in animal, wet weight/concentration in soil, wet weight or dry weight.

al.[31] reveal little increase in concentration in foliage over soil. This result is due to the desorption of compounds from soil yielding low concentrations in air, which are then accumulated in foliage to achieve concentrations somewhat similar to those originally present in the soil. However, if the bioaccumulation factors were expressed in terms of atmospheric concentration, much higher levels of BF would be expected.

2. Soil-to-Animal Process

Agricultural pesticides are broadcast over crops and substantial residues can accumulate in soils in these treated areas. In this way, earthworms, snails, slugs, various insects, and other soil invertebrates are frequently exposed to lipophilic xenobiotic compounds. A number of investigations have been conducted into the accumulation of soil residues of mainly lipophilic agricultural pesticides by soil invertebrates. Some of the data obtained is summarized in Table 5.

Probably, most data and information is available on the soil-to-earthworm transfer. In investigation of DDT isomers and DDE, Davis and French[37] found that a major factor influencing the amount of bioaccumulation in the worms was the concentration in the soil. Similarly, but with a wider range of chlorinated hydrocarbon insecticides, Wheatley and Hardman[35] observed the same relationship as shown in Figure 8. The regression equation observed was

$$\log C_{wo} = 0.27 + 0.80 \log C_s$$

where C_{wo} is the worm concentration and C_s the soil concentration. This relationship accounts for 95% of the variation observed.

In later investigation of DDT, deildrin, and heptachlor by Gish and Hughes,[38] the following equation accounted for 80% of the variability:

$$\log C_{wo} = -1.62 + 0.66 \log C_s - 0.03 \, T + 0.04 \, M_{ow}$$
$$+ 0.07 \, I - 0.36 \, L - 0.02 \, M_{os}$$

where C_{wo} is the worm concentration (ppm), C_s is the soil concentration (ppm), T is the time period after treatment (months), M_{ow} is the percent moisture in the worms, I is the inches of rainfall in the previous two weeks, L is the percent lipid in the worms, and M_{os} is the percent moisture in the soil.

This indicates that the residue concentration in the soil and the lipid content of the worms are the most important factors, and the other factors have a much lower level of importance. It is interesting to note that the organic content of the soil was found to have little influence on the relationship outlined above.

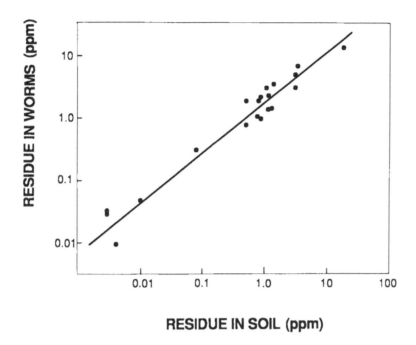

RESIDUE IN SOIL (ppm)

FIGURE 8. Relationship between the concentration of organochlorine in earthworms and the associated soil, calculated from 90 paired values. (From Wheatley, G. A. and Hardman, J. A., *J. Sci. Food Agric.*, 19, 219, 1968. Copyright Blackwell Scientific Publications. With permission.)

The mechanisms of accumulation of lipophilic compounds have been investigated by Lord et al.[39] These authors have suggested that the bioaccumulation process could be seen as a soil-to-soil water equilibrium followed by a soil-water-to-worm equilibrium. The soil-to-soil-water distribution was found to be represented by the following equation

$$\log K_{sw} = 0.48 \log K_{ow} + 1.04 \tag{2}$$

where K_{sw} is the equilibrium partition coefficient between soil and soil water.

In addition, the worm body solids were equilibrated with water and the process found to be represented by the following equation:

$$\log K_{ww} = 0.53 \log K_{ow} + 0.69 \qquad (r = 0.95) \tag{3}$$

3. Soil-to-Earthworm Process — A Proposal

Lord et al.[39] have suggested that the addition of the equations above would indicate that BF for all compounds would fall into a narrow range and be influenced by the soil organic matter content. Also, investigations of the kinetics of these processes gave indications of the time periods required.

The process, as outlined by Lord et al.[39] as above, can be generalized and simplified, as shown diagrammatically in Figure 9. This is somewhat analogous to the uptake of lipophilic compounds by aquatic infauna, as described in Chapter 6. Utilizing this approach and first considering the soil to soil water equilibrium

$$K_{sw} = f_{oc} K_{oc}$$

but

$$K_{oc} = x K_{ow}^{a}$$

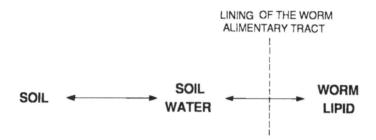

FIGURE 9.   Proposed pathway of bioaccumulation of lipophilic compounds from soil by earthworms.

where F_{OC} is the fraction of organic carbon present in the soil, K_{OC} is the sediment-to-sediment water partition coefficient in terms of organic carbon, x is a proportionality constant, and constant a is a nonlinearity constant. Therefore

$$K_{SW} = x \ f_{oc} \ K_{OW}^a \tag{4}$$

but generally, as discussed in Chapter 6, for uptake of lipophilic compounds by organisms from water the following relationship applies:

$$K_B = y_L \ K_{OW}^b \tag{5}$$

where K_B is the bioaccumulation concentration factor in the organism-to-water system, y_L is the lipid content of the organism, and constant b is a nonlinearity constant.

$$\text{Now} \quad BF = C_{WO}/C_S = (C_{WO}/C_W)(C_W/C_S)$$

where BF is the bioaccumulation factor and C_W the concentration of chemical in water, thus

$$BF = K_B/K_{SW} \tag{6}$$

and from Equations (4) and (5), then

$$BF = y_L \ K_{OW}^b/x \ f_{oc} \ K_{OW}^d = x(y_L/f_{oc}) \ K_{OW}^{b-a} \tag{7}$$

$$\text{and} \quad C_{WO} = x(y_L/f_{oc}) \ K_{OW}^{b-a} \ C_S$$

Utilizing the relationships produced by the experimental work of Lord et al.[39] then from Equation 2

$$K_{SW} = 11 \ K_{OW}^{0.48}$$

and from Equation 3

$$K_{WW} \text{ (equivalent to } K_B) = 4.9 \ K_{OW}^{0.53}$$

thus from Equation 6

$$BF = (4.9/11) K_{OW}^{0.05}$$

This suggests that BF in this system will have a weak dependence on the K_{OW} value of a compound. If other values, mentioned in Chapter 6 previously, are used for constants a and b, there remains a weak or negligible dependence of BF on K_{OW}, since constants a and b both usually approach unity.

This lack of dependence on values of K_{OW} is implied in the results of many authors, since such a dependence has not yet been reported. This means that BF values for all lipophilic compounds will be similar. This is in accord with BF values reported for earthworms in Table 5, which cover the range 0.12 to 19.5. Usually, processes which are strongly dependent on the K_{OW} values of the compounds involved exhibit a range of bioaccumulation values of several orders of magnitude.

According to Equation 7 the major influence on BF will be the ratio of the lipid content in the worms to the organic carbon content in the soil. This general equation for BF suggests that this characteristic will be directly related to the lipid content of the worms and inversely related to the organic carbon content in the soil.

Let us now turn to considerations of the factors which have an influence on the concentrations which bioaccumulate in worms. Gish and Hughes[38] found a strong dependence of the worm concentration on the worm lipid content, but the soil organic carbon content influence was negligible. On the other hand, Lord et al.[39] have provided evidence of an inverse relationship between worm concentration and soil organic carbon content, and this has also been observed by Davis.[40] Since K_{OW}^{b-a} is usually small or negligible, and x, y_L and f_{OC} are constants, the concentrations which are bioaccumulated in the worms will be almost directly proportional to the soil concentration. This has been observed by a number of authors and is illustrated in Figure 8.

The validity of this general approach, as outlined above, is suggested by the current information and data available, but it requires verification. There are a large number of other factors which may exert an influence and modify the relationships which have been suggested. For example, the approach outlined above assumes equilibrium, but this may not be attained in all situations.

Residues of persistent agricultural chemicals, such as chlorinated hydrocarbon insecticides, have been observed in the muscle and other tissues of grazing animals such as beef cattle. Often these substances originate from residues in the soil. Several routes from soil to grazing animals are possible, one of which is ingestion of soil during feeding and transfer of the pesticide to lipid tissues in the animals by mechanisms similar to those described for earthworms.

C. BIOMAGNIFICATION IN AQUATIC SYSTEMS
1. Routes for Biomagnification

Previously, in Chapters 4 and 6, bioconcentration and its relationship to biomagnification were described, and this was illustrated diagrammatically as shown in Chapter 6, Figure 1. Biomagnification can only apply to aquatic organisms which obtain xenobiotic compounds through residues in food. Thus, this mechanism only relates to higher trophic level organisms which consume complex organic matter as food. This includes organisms such as fish, crabs, many invertebrates, and so on. Of course, these biological classes can only be seen as general guidelines for trophic status, and to be certain of uptake routes for an organism, it must be subject to an appropriate investigation. In addition, the organism must be aquatic to the extent that equilibrium can be established with the surrounding water.

A diagrammatic illustration of the pathways of bioaccumulation of a xenobiotic chemical is shown in Figure 10. Biomagnification is the transfer of a chemical from food to organism, but clearly cannot occur without the movement of the substance from the gastrointestinal

AQUATIC ORGANISM

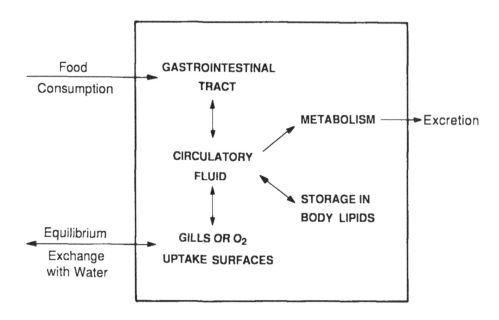

FIGURE 10. A diagrammatic illustration of the patterns of bioaccumulation of a xenobiotic chemical in an aquatic organism.

tract to the circulatory fluid, and then through the gills to the surrounding water. So the clearance process involved in bioconcentration is also involved in biomagnification. However, different situations may apply with the chemical in the surrounding water with regard to its concentration. The concentration can range from zero to higher values, and if the substance is present at all in water, then bioconcentration, at least to some extent, will occur at the same time as biomagnification.

The biomagnification process involves transfer of the xenobiotic lipophilic chemical from the gastrointestinal tract through its walls to the circulatory fluid, as shown in Figure 10. The circulatory fluid comes in contact with most body tissues, and the lipophilic compounds are deposited in fatty tissues.[41-43]

There is a large volume of data on the transfer efficiency of this process for a variety of compounds and organisms, and some examples are shown in Table 6. Little relationship can be found between transfer efficiency and the physiochemical and other properties, although Ellgehausen et al.[56] found a relationship between log K_{ow} and transfer efficiency. It is interesting to note that after experiments on rats, Burgman et al.[52] reported that PCBs had transfer efficiencies from 94 to 28%, and this appeared to decline with increasing numbers of chlorine atoms in the molecule. It is also significant that Muir and Yarechewski[53] have noted a marked decline in efficiency with fish with compounds having log K_{ow} values greater than 6. In fact, these exhibit efficiencies close to 0 at log K_{ow} values greater than 8.5. These compounds fall into the superhydrophobic category, and this behavior is consistent with previous discussion regarding the properties of this class in Chapter 6, Section II.A.2.

Many factors could be expected to influence the transfer efficiency. Harding et al.[50] have described a number of factors which affect the feeding rate of copepods, and these could be expected to have some effect on the transfer efficiency. These factors include temperature, type of food, food size, and so on. These factors may generally influence the transfer efficiency with all aquatic organisms. However, one factor which has been clearly demonstrated to be important in transfer efficiency is the ingestion rate, as illustrated in

TABLE 6
Examples of Transfer Efficiencies and Biomagnification Factors for One-Step Biomagnification

Substance	Food	Consumer	Transfer efficiency (%)	Biomagnification factor (BF)[a]	Ref.
HEOD	Tuberficid worms	Reticulate sculpin	About 80	—	44
DDT	Pelleted fish chow	Goldfish	—	0.078 (muscle)	42
DDT and dieldrin	Prepared fish food	Rainbow trout	DDT 20—24 Dieldrin 9—11	—	43
PCB	Control diet	Rainbow trout	68	0.57	45
Kepone	Algae	Oysters	—	0.007	46
	Brine shrimp	Mysids	—	0.5	
	Mysids	Spot	—	0.85	
DDT	Clams	Flathead minnows	—	1.2	92
B(a)P	Diatom	Laval bivalve	5.4	0.42	47
DDT, Fluorodifen Terbu- tryn Atrazine	Algae	Daphnids	3.9—23	—	48
PCB	Daphnids	Catfish	9.1—44	—	48
	Dry fish food	Goldfish	>40	0.2—1.7 (lipid)	49
DDT	Phytoplankton	Copepod	10—70	—	50
Kepone	Daphnids	Bluegills	Mirex 34 Kepone 14	Mirex 1.1 Kepone 0.46	51

[a] BF = concentration in consumer/concentration in food.

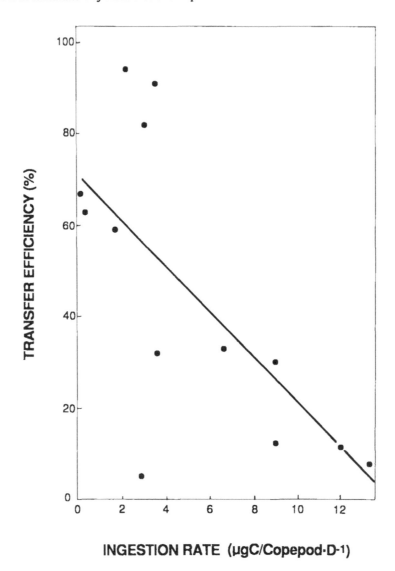

FIGURE 11.   The relationship between transfer efficiency of DDT in food to the copepod, *Calanus sinmarchicus,* and ingestion rate of food. (From Harding, G. C., Vass, W. P., and Drinkwater, K. F., *Can. J. Fish. Aquatic Sci.,* 38, 101, 1981. Copyright Government of Canada. With permission.)

Figure 11. Here the transfer efficiency of DDT in food of copepods declines with increasing ingestion rate.

Similar to other bioaccumulation processes, the compounds which exhibit biomagnification are usually resistant to degradation and exhibit the general properties described in Chapter 3. Thus, compounds exhibiting a significant degree of biodegradation over the biomagnification period will exhibit correspondingly less biomagnification.

2. The Biomagnification Factor (BF) — A Proposal

The Biomagnification Factor (BF $= C_C/C_F$ where C_C is concentration in consumer and C_F is concentration in food) has been determined for a wide range of different food-to-consumer situations. A selection of a number of these values is shown in Table 6. In contrast to the Bioconcentration Factors for similar compounds, which exhibit values ranging over

TABLE 7
Values for the Biomagnification Factor (BF)ᵃ for Salmon on Different Diets Containing Various Compounds and Carried Out Under Consistent Conditions

Compound	Biomagnification factor					
	Diet 1	Diet 2	Diet 3	Diet 4	Diet 5	Average
PCBs	0.87	0.91	1.08	1.49	1.00	0.87
HCB	0.40	1.32	0.69	1.03	—	0.86
α BHC	0.12	—	—	—	—	—
Lindane	0.36	—	—	—	—	—
DDE	0.68	0.70	1.09	1.43	1.72	1.12
o,p — DDT	1.60	0.75	1.09	—	—	1.14
p,p — DDT	1.61	0.51	0.81	—	—	0.98
DDD	1.18	0.71	0.95	—	1.21	1.01
α Chlordane	0.81	0.67	0.71	0.95	0.80	0.79
Chlordane	0.40	0.17	0.69	—	—	0.42
Minex	1.11	—	—	—	—	—
Average	0.88	0.84	1.06	1.50	1.06	—

ᵃ Wet weight basis.

Adapted from Leatherhead, J. F. and Sonstegard, R. A., *Comp. Biochem. Physiol.*, 72C, 91, 1982.

TABLE 8
Values for the Biomagnification Factor (BF)ᵃ for Trout on Different Diets Containing Various Compounds and Carried Out Under Consistent Conditions

Compound	Biomagnification factor					
	Diet 1	Diet 2	Diet 3	Diet 4	Diet 5	Average
Σ DDT	2.20 (0.9)	2.1 (1.0)	2.3 (1.0)	3.8 (1.7)	2.6 (1.1)	1.1
Chlordane	3.1 (1.3)	2.1 (1.0)	5.0 (2.2)	7.3 (3.4)	2.0 (0.9)	1.8
Dieldrin	3.0 (1.3)	1.3 (0.6)	2.2 (1.0)	2.9 (1.3)	2.2 (1.0)	1.0
PCB	2.2 (0.9)	2.7 (0.6)	3.5 (1.5)	4.1 (1.9)	2.1 (0.9)	1.2
Average (lipid basis)	0.9	0.8	1.4	2.1	1.0	

ᵃ Calculated on a wet weight basis with the figure in parenthesis being BF on a lipid weight basis.

Adapted from Hilton, J. W., Hodson, P. V., Braun, H. E., Leatherhead, J. L., and Slinger, S. H., *Can. J. Fish. Aquatic Sci.*, 40, 1987, 1983.

several orders of magnitude, these values exhibit a limited range, i.e., between 0.2 and 1.7, as illustrated by the data in Table 6. Also, the values for BF are very small, i.e., 0.2 to 1.7, compared to the values for the Bioconcentration Factor of similar compounds, which range from 10^2 to 10^6.

Some investigations have been carried out on the biomagnification of a range of compounds in the diets of fish maintained under consistent conditions. This data allows a more valid direct comparison to be made of BF values of different compounds. The experiments have been repeated with different diets containing the same compounds and yielding the results shown in Tables 7 and 8. A number of conclusions can be drawn based on these results. First, all of the compounds in Table 7 have a similar biomagnification factor. Second,

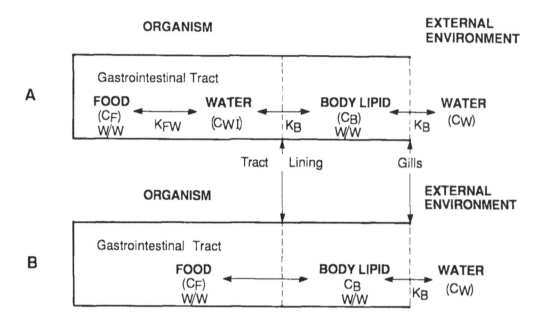

FIGURE 12. Two exchange processes involved in the biomagnification of persistent lipophilic compounds in food by aquatic biota.

there is no indication of a relationship between BF and K_{OW} with the data in either Table 7 or 8. In the case of BF values expressed on a wet weight basis in Table 7, the diets used were in fact based on salmon as both the food and the consumer. Thus, the lipid content in the diet and the fish would be expected to be comparable, and the BF values obtained similar to those values for BF calculated on a lipid basis. All of the BF values in Tables 7 and 8, on a lipid basis extend over a narrow range, i.e., 0.42 to 1.8, and have an average value of 1.02.

These factors can be explained by the following simple interpretation which only takes into account partition processes between the food and the consumer. The bioaccumulation of persistent lipophilic compounds from food by aquatic biota can be seen as proceeding by two routes, as shown in Figure 12. One route (A, in Figure 12) is by exchange of compounds in food with water in the gastrointestinal tract, and then exchange of these mobilized compounds through the walls of the tract with body lipids by the circulatory fluid. The other route (B in Figure 12) is by direct exchange of compounds in food through the walls of the gastrointestinal tract with the circulatory fluids and the body lipids.

First, consider route A and the food-to-gastrointestinal tract water exchange as the initial process involved. Even though food is moving through the tract, it can be seen as establishing an equilibrium with the water present. The food is subject to movement and digestive processes which should facilitate this equilibrium. The equilibrium can be seen to be analogous to the organism-to-water equilibrium outlined in Chapter 6, or the blended worm-to-water equilibrium investigated by Lord et al.[39] Accordingly, it is comparable to bioconcentration and should be governed as similar factors at equilibrium. However, the kinetics of this process would be expected to be entirely different. Thus

$$K_{FW} = y_{LF} K_{OW}^b \tag{8}$$

where K_{FW} is the partition coefficient between water and food, y_{LF} is the lipid fraction in the food, and constant b is a nonlinearity constant.

Turning to the next step in the process, the water-to-organism equilibrium, this is similar to the bioconcentration process at equilibrium, thus

$$K_B = y_{LB} K_{OW}^a \qquad (9)$$

where K_B is the bioconcentration factor, y_{LB} is the lipid content of the biota, and constant a is a nonlinearity constant.

It should be kept in mind that in this case also, the kinetics of the process may differ from the body lipid-to-external-water process. Now

$$BF = C_B/C_F = (C_B/C_{WI})/(C_{WI}/C_F) = K_B/K_{FW}$$

where C_B is concentration in the organisms, wet weight; C_F is concentration in food, wet weight; C_{WI} is the concentration in water in the gastrointestinal tract. Substituting Equations (8) and (9) for K_B and K_{FW} into this equation, then

$$BF = (y_{LB} K_{OW}^a)/(y_{LF} K_{OW}^b) = (y_{LB}/y_{LF}) K_{OW}^{(a-b)} \qquad (10)$$

But as discussed in Chapter 6, for most bioconcentration processes constant b approaches unity. The food-to-water process is very similar to this, and a constant a value of approximately unity could be expected here also. Thus

$$BF = y_{LB}/y_{LF}$$

$$\text{and} \qquad C_B = (y_{LB}/y_{LF})C_F$$

Considering the biomagnification route B shown in Figure 12, this path is essentially equivalent to the direct diffusion of compounds from food lipid to body lipid. When equilibrium is established the concentration of the lipophilic compound in both food and body lipids will be the same. Thus

$$BF = C_B/C_F = y_{LB}/y_{LF} \qquad (11)$$

and it can be shown that

$$(C_B/y_{LB})(y_{LF}/C_F) = 1 \qquad \text{and}$$

$$C_{BL}/C_{FL} = 1$$

where C_{BL} and C_{FL} are the concentrations expressed in terms of lipid weight. Also

$$C_B = (y_{LB}/y_{LF})C_F$$

This means that both paths lead to the same relationships at equilibrium, but the kinetics of these processes are probably quite different.

In the previous section a variety of factors were mentioned which could affect the transfer efficiency of compounds from food to the organism. This included such aspects as feeding rate, temperature, and so on. While these factors may affect the transfer efficiency, it is likely that they do not affect the final equilibrium between food and organism. However, they would be expected to strongly influence the kinetics of the various equilibria discussed above, and thus the time required to achieve equilibrium.

The previous discussion indicates that while the rate constants for the various equilibria are constant, the actual rate of movement of a compound through the system will change with time of exposure. At the early stages of exposure, movement from food to body lipid would be rapid, but slowing as equilibrium is approached. This suggests that the transfer efficiencies will also change with time period of exposure and differ for the same bioconcentration process, depending on the time period over which the measurement is made.

The final process involved in biomagnification with aquatic organisms is the movement through the circulatory fluid and gills, or other oxygen extraction organs, to the external water, and the reverse movement. If overall equilibrium between all parts of the system is established, then concentrations in food, water in the gastrointestinal tract, body lipid, and external water will stabilize, and the equations previously derived will apply. However, if the external water concentration approaches zero and remains at that value, then there will be a slow rate of compound through this route, but this would probably be relatively small and can be neglected.

The relationships derived above for BF suggest that, provided equilibrium is reached:

1.   BF will be effectively independent of K_{ow} or weakly dependent on K_{ow}.
2.   BF calculated using concentrations for organisms and food expressed in lipid weight terms will be approximately unity for all lipophilic compounds.
3.   The concentration in the organism will be directly proportional to the concentration in food.
4.   BF in wet weight terms will be the ratio of the proportions of lipid in organism and food.
5.   BF will be directly proportional to the lipid content of the biota, and inversely proportional to the lipid content of food.

The available results typified by the data in Tables 6, 7, and 8 lend general support to the validity of items 1 and 2 above, as discussed previously. Both Hilton et al.[54] and Leatherland and Sonstegard[55] have reported that in their experiments on biomagnification with salmon the organism concentration was directly proportional to the food concentration, as suggested in item 3 above. The remaining items require further experimental investigation. Compounds which degrade over the biomagnification period would be expected to exhibit reduced concentration in the organism and also reduced biomagnification factors.

While the mechanisms previously proposed are generally supported by the results available, there are a variety of factors which need further investigation and validation. For example, the nature of the equilibrium between food and water needs clarification, particularly in terms of the continuous movement of food through the gastrointestinal tract. Also, the kinetics of the various equilibria would be expected to play an important role in the processes involved, as do these factors in the bioconcentration process.

3. Biomagnification Factors (BF) in Aquatic Food Chains — A Proposal

Early data on the residues of persistent pesticides in food chains, including aquatic food chains, suggested that there was an increasing concentration of these residues with increasing trophic level. Rudd[57] prepared an outline of much of this evidence in 1964; examples of this are shown in Figure 13. Similarly, Woodwell[58] assembled data on concentrations in biota worldwide in 1967; this data is shown in Table 4 in Chapter 4. The explanation of the mechanism was that removal of food material by respiration resulted in concentration of the remaining xenobiotic biodegradation-resistant substances. This mechanism is also discussed in Chapter 4. Since that time further evidence has suggested that bioconcentration, the water-to-organism process, is the major controlling mechanism for bioaccumulation of the xenobiotic chemicals in aquatic organisms. However, there remains no adequate me-

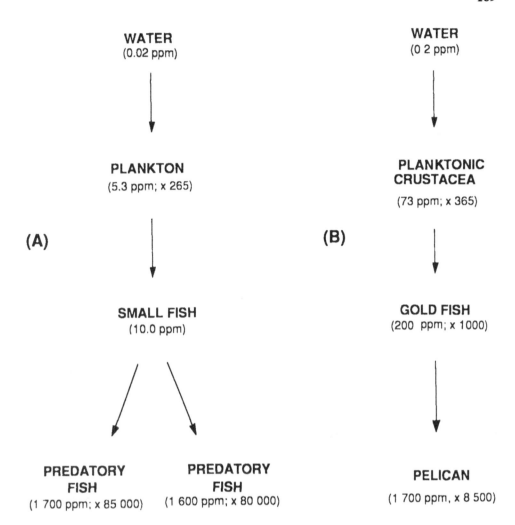

FIGURE 13. The biomagnification of insecticides in aquatic food chains with (A) DDD and (B) toxaphene. (Adapted from Rudd, R. L., *Pesticides and the Living Landscape*, University of Wisconsin Press, Madison, 1964.)

chanistic explanation of biomagnification in aquatic food chains, which is an important bioaccumulation process. The following proposal provides a reasonably acceptable explanation for much of the data on behavior of persistent lipophilic compounds in food chains. However, such a simple explanation must result in inadequate consideration of many factors which may prove to be of considerable significance in some situations.

In the previous section a proposal was made that the existing one-step biomagnification factor data could be substantially explained by a mechanism involving partitioning between food and gastrointestinal water, and body lipids or equilibrium movement between the lipids, in the consumer and those in the food. The situation in food chains involving a sequence of one-step transfers from food to consumer was not considered. The transfer of a chemical in an aquatic system, incorporating a food chain, can be diagrammatically represented, as in Figure 14. Entry of chemicals to any organisms in the food chain is through two sources: first, the food, and second, exchange with chemicals in the external ambient water. However, the chemical in food initially originates from the first organism in the food chain, in this case the phytoplankton.

In artificial systems, such as some of those considered in the previous section, the concentration in the food was often arbitrary. In aquatic ecosystems at equilibrium, the only

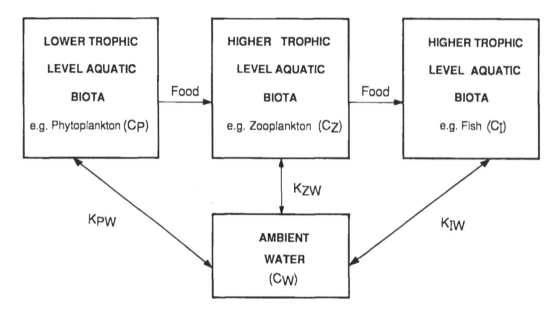

FIGURE 14.   The transfer of persistent chemical in an aquatic food chain.

source of xenobiotic chemical for the lower trophic level biota, e.g., phytoplankton, is from the ambient water. Since these organisms are autotrophic, they do not utilize complex food substances which could contain residues of xenobiotic chemicals. This means the first step in an aquatic food chain is always a bioconcentration step, and can be represented by Equation 9, as discussed in Chapter 6. Thus

$$K_{PW} = C_P/C_W = y_{LP} K_{OW}^a$$

where K_{PW} is the bioconcentration factor from water-to-lower trophic level aquatic biota, C_P is the lower trophic level biotic concentration, C_W is the water concentration, y_{LP} is the lipid concentration in the lower trophic level biota, K_{OW} is the octanol-to-water partition coefficient of the compound involved, and constant a is a nonlinearity constant. Since the concentrations involved in this process are comparatively low, and from the discussion of the overall data in Chapter 6, it can be assumed that the constant a is effectively unity. Thus

$$K_{PW} = y_{LP} K_{OW}$$

$$\text{and} \qquad C_P = y_{LP} K_{OW} C_W \qquad (12)$$

Now considering the second step in the food chain, the bioaccumulation of chemical by higher trophic level aquatic biota, the bioconcentration of chemical from water alone, by reasoning similar to that above, can be represented by

$$C_Z = y_{LZ} K_{OW} C_W$$

where C_Z is the concentration of the chemical in higher trophic level biota, and y_{LZ} is the lipid proportion in the higher trophic level biota.

Let us turn to the uptake of chemicals from food alone by the higher trophic level aquatic biota in this second step. In this situation, utilizing Equation 11 derived in the previous section, then

$$BF_{ZP} = y_{LZ}/y_{LP}$$

where BF_{ZP} is the biomagnification factor (C_Z/C_P), thus

$$C_Z = (y_{LZ}/y_{LP})\, C_P$$

but substituting for C_P from Equation 12

$$C_Z = (y_{LZ}/y_{LP})\, y_{LP}\, K_{OW}\, C_W = y_{LZ}\, K_{OW}\, C_W$$

This result indicates that irrespective of whether the uptake route is food or water, the relationship between the concentration in the organism and the concentration in ambient water is the same and equivalent to that obtained by bioconcentration alone.

This procedure can be repeated with other members of the food chain and in the general case

$$C_I = y_{LI}\, K_{OW}\, C_W$$

and

$$BF_I = C_{I1}/C_{I2} = y_{LI1}/y_{LI2}$$

where C_{I1} and C_{I2}, as well as y_{LI1} and y_{LI2}, are the concentrations of chemical and lipid in two sequential members of an aquatic food chain. Also, by reasoning similar to that used in the previous section, the BF_I expressed in terms of lipid weight would be unity in all cases.

This treatment indicates that all of the five suggestions listed in the previous section in relation to factors influencing the BF may be applicable to all members of an aquatic food chain. In addition to these, the following can be suggested to apply, provided equilibrium is attained:

1. The bioconcentration factor (C_B/C_W) for each member of a food chain will be similar to that obtained by considering water alone as the source of chemical, irrespective of whether food or water is the actual source.
2. All members of the food chain should have BFs which approach unity if concentrations are expressed in terms of lipid weight.
3. The actual concentrations which occur in any organism will be directly proportional to the organism lipid content.

It is now appropriate to evaluate the relationship of these suggestions to the existing data. First, considering the data in Figure 13, the water-to-plankton step and the observed bioconcentration factors are approximately in accord with the mechanism outlined above. The residue concentrations of 5.3 and 10.0 ppm DDD in plankton and fish, respectively, could be explained by differences in lipid content, as could the concentrations of 73 and 200 ppm toxaphene in planktonic crustaceans and goldfish, respectively. But a concentration of 1700 ppm DDD in predatory fish is difficult to reconcile with the proposed mechanism. Considering the data by Woodwell[58] in Table 4 in Chapter 4, the fully aquatic organisms show little evidence of food chain increases in residue concentrations. The differences which are observed could be due to different fat contents, and perhaps different exposures to the chemical. Birds and mammals are not fully aquatic organisms, and will be considered in more detail in this Chapter, Section III.D. on Biomagnification in Terrestrial Systems.

A variety of other information is available from more recent investigations. For example, Bahner et al.,[46] in a controlled food chain experiment, found that the biomagnification factors for kepone in transfers from spot to mysid and mysid to brine shrimp were essentially equal at 0.585 and 0.53. These values were calculated in terms of wet weight and could approach unity if calculated in terms of lipid weight. In another investigation of kepone by Bender and Huggett,[59] the concentration of residue in oysters, blue crabs, perch, and the American eel were found to be 0.06, 0.19, 0.34, and 0.32 mg kg^{-1} wet weight, respectively, after a period of several years. These are somewhat similar and when adjusted for lipid content could be expected to approach more closely the same numerical value. Also, in investigations of PCBs in an estuarine ecosystem, Shaw and Connell[60] found no evidence of increases in residue with position in the food chain, or trophic status. In fact, the concentrations in fish, mussels, polychaetes, and crabs occur over a relatively narrow range, and are probably explicable in terms of exposure and lipid content. There are many sets of data in the literature which show no evidence of food chain biomagnification in fully aquatic systems. Also, Gunkel and Streit[61] have described in some detail the exchange process between water, blood, muscle, and so on for bioaccumulation of atrazine by molluscs and fish. The mechanisms derived by these authors are somewhat similar to the processes described above. Most of this data tends to support the mechanism previously outlined, but more specific data will be needed to provide unequivocal evidence. Factors such as growth of the organisms, rates of uptake and clearance, feeding behavior, and so on are not taken into account in this approach and would be expected to exert an influence.

4. Models for Biomagnification in Aquatic Systems

In the previous two sections on the biomagnification factor equilibrium is assumed, and little can be learned in this approach regarding the kinetics of the processes involved. The following models generally attempt to incorporate the rates at which various processes occur, as well as other factors. Few of these models, however, incorporate the equilibrium processes previously outlined, and some are based on assumptions shown in later work to be inaccurate or quite incorrect. Of course, it should be emphasized that these latter models cannot yield accurate results, although in some cases they may produce information that agrees with a limited data set.

In 1971 Eberhardt et al.[62] produced one of the first models for behavior of a persistent lipophilic compound in the environment. DDT in a freshwater marsh was modeled based on first order kinetics and fast and slow components within organisms. Several years later, Isaacs[63] developed a model based on unstructured food webs to explain aspects of biomagnification in food webs. A model based on pollutant biokinetics and fish energetics was developed by Norstrom et al.[64] Both uptake in food and water were taken into account, and clearance was based on body weight. The model successfully predicted concentrations of PCBs and methylmercury in yellow perch in the Ottawa River. Fagerstrom[65] and Thomann[66] have also used body weight as a basis for modeling trace substance behavior in aquatic organisms.

Differences in PCB concentrations between five species of fish in lakes have been explained by Jensen et al.[67] using a bioenergetic model. This model describes uptake in terms of metabolism, food consumption, size, and growth. Thomann and Connolly[68] have developed a bioenergetic model which takes into account exposure through food and water, which agrees with field data on Lake Trout. The model predicted that food chain sources of PCB would account for 99% of these residues in this top predator. It is interesting to note that predictions of fish residue concentrations, assuming bioconcentration and using the K_{OW} to K_B relationships previously described in Chapter 6, were incorrect by a factor of about 4.

Extensive laboratory-based experiments have been carried out on biomagnification and

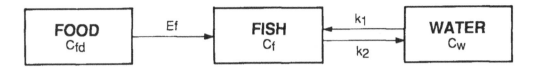

FIGURE 15. The transfer processes involved in biomagnification with fish, where C_f, C_{fd} and C_w are the concentrations in food, fish, and water, respectively; E is the absorption efficiency for ingested chemical; f is the feeding rate, and k_1 and k_2 are the uptake and clearance ratio constants from water. (From Bruggermann, W. A., Marton, L. B. J. M., Kooiman, D., and Hutzinger, O., *Chemosphere*, 10, 811, 1981. Copyright Pergamon Press. With permission.)

bioconcentration by Bruggeman et al.[49] The model outlined in Figure 15 was used as a basis for prediction of behavior patterns of chemicals. The basic relationship was expressed by

$$D\ C_f/dt\ =\ Ef\ C_{fd}\ +\ k_1\ C_w\ -\ k_2\ C_f$$

where C_f, C_{fb}, and C_w are concentrations in fish, food, and water, respectively, E is the absorption efficiency for ingested chemical, f is the feeding rate, and k_1 and k_2 are the uptake and clearance rate constants to and from water.

A number of different environmental situations were investigated using this equation by the authors. With constant dietary exposure at equilibrium

$$BF\ =\ Ef/k_2$$

They concluded that the food chain was likely to be an important source of xenobiotic chemicals only with persistent chemicals having extremely low water solubility.

In another approach, Connolly and Pedersen[69] investigated both a thermodynamically-based (fugacity model) and a kinetic bioenergetic approach (food chain model) of the behavior of chemicals in aquatic food chains. Using laboratory and field data, these authors found that the ratios of fugacities in field animals and water were generally above unity and increased with trophic level and the hydrophobicity of the chemical. This is consistent with the food chain model, but violates the assumption of the fugacity model that the maximum fugacity ratio is unity.

D. BIOMAGNIFICATION IN TERRESTRIAL SYSTEMS

1. Routes for Biomagnification in Terrestrial Organisms

Biomagnification is the process of transfer of xenobiotic lipophilic chemicals from food to terrestrial organisms, often resulting in an increase in the concentration in the consumer over that in the food. Examples of some terrestrial organisms involved in this process are outlined in Table 1. The organisms involved are terrestrial animals, including insects and other appropriate invertebrates, which consume complex food materials which can contain residues of xenobiotic chemicals. In this context, aquatic organisms are those which establish a direct equilibrium partition relationship with chemicals in the water mass. Terrestrial organisms are those that do not have this relationship, and thus include such animals as porpoises, whales, aquatic birds, and so on. So terrestrial organisms include mammals, birds, many insects and invertebrates, and so on, and are included in the categories of lower trophic level terrestrial biota and higher trophic level terrestrial biota shown in Figure 1.

Biomagnification with terrestrial biota by paths other than food have been previously considered in this chapter, Sections III.A.2. and III.B.2., where uptake from the atmosphere and soil by terrestrial biota were considered. These routes and the food pathways differ in their importance, depending on the xenobiotic chemicals involved, the organism, environmental factors, and so on. So to evaluate biomagnification with terrestrial organisms a knowledge of the routes of uptake is required.

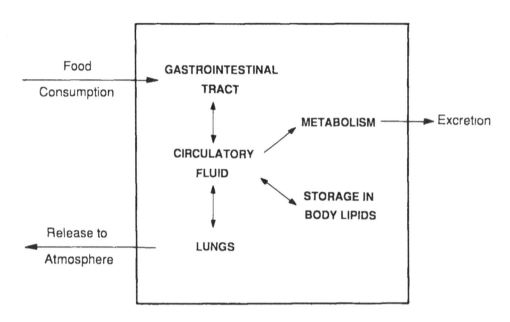

FIGURE 16.   A diagrammatic illustration of the patterns of biomagnification of a xenobiotic lipophilic chemical in a terrestrial organism.

A simplified illustration of the behavior of lipophilic chemicals on biomagnification from food by a terrestrial organism is shown in Figure 16. The chemical in food passes through the walls of the gastrointestinal tract into the circulatory fluid, then to organs where it can be metabolized, or is deposited in body lipids. Metabolism usually involves conversion of the lipophilic compound to a water soluble compound by the attachment of polar, usually oxygen-containing, groups. The metabolite can then be excreted in aqueous solution.

Movement of the compound in circulatory fluid to the atmosphere can occur through the lungs. Due to the size and movement of the atmosphere, under normal circumstances it is unlikely that an increase in the atmospheric concentration of xenobiotic chemical would occur, and so equilibrium would not normally be established. But in some situations, where it is possible for the concentration in the atmosphere to increase, equilibrium is possible. This route of clearance of compound is probably important only for compounds with relatively high vapor pressures. With most xenobiotic chemicals, clearance by this route is probably relatively slow compared with metabolism and excretion.

There is a reasonable body of evidence available to support the biomagnification mechanism outlined in Figure 16 for terrestrial organisms. The first step in this mechanism is the establishment of an equilibrium between food in the gastrointestinal tract and the circulatory fluid. This equilibrium has been investigated with birds by Henny and Meeker,[70] who found the relationship shown in Figure 17 for DDE. The equilibrium constant for this relationship, 0.042, would be expected to be related to the lipid content of the diet and that of the circulatory fluid (plasma).

The circulatory fluid (plasma)-to-body-lipid equilibrium has been investigated with a number of different organisms. For example, Wolff et al.[71] found the relationship shown in Figure 18 with human subjects exposed to PCBs. Henny and Meeker[70] determined that the plasma DDE concentrations were correlated with brain concentrations and egg concentrations with birds. Brown and Lawton[72] found that there was a simple relationship for partition of PCBs within the human body at equilibrium. The adipose tissue to serum PCB concentrations

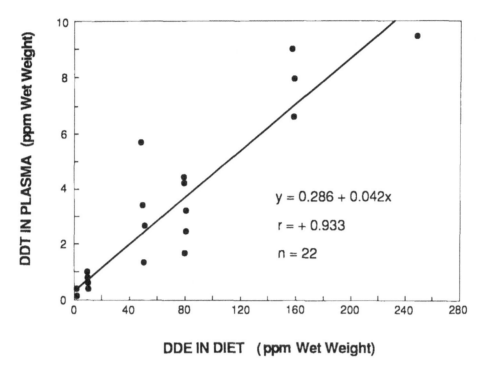

FIGURE 17. The relationship between DDE in the diet (wet weight) and DDE in blood plasma (wet weight) with American kestrels. (From Henny, C. T. and Meeker, D. L., *Environ. Pollut.* (Series A), 25, 291, 1981. Copyright Applied Science Publisher. With permission.)

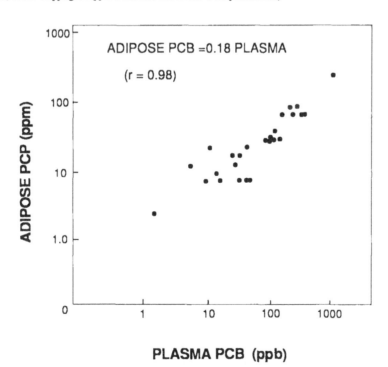

FIGURE 18. Relationship between PCB concentration in adipose tissue and plasma (both wet weight) with human subjects exposed to PCBs. (From Wolfe, M. S., Thornton, J., Fischbein, A., Lilis, R., and Selikoff, I. J., *Toxicol. Appl. Pharmacol.*, 62, 294, 1982. Copyright Academic Press. With permission.)

were related to their lipid contents. Consistent with this, Kan and Rooyen,[73] using hens, have found high correlations between the concentrations of a variety of organochlorine residues in various tissues and fat. Somewhat similar relationships have been found by Tanabe et al.[74] with penguins, Norstrom et al.[75] with herring gulls, and Kan and Rooyen[76] and Forsyth and Peterle[77] with birds.

2. The Biomagnification Factor

The biomagnification mechanism, as outlined in the previous section, is similar to that proposed for biomagnification with aquatic organisms (this chapter, Section III.C.2.). By reasonsing similar to that used previously with Equation 10 it can be shown that

$$BF = (y_{LB}/y_{LF})K_{OW}^{a-b} \tag{13}$$

where BF is the biomagnification factor (C_B/C_F with C_B as concentration in biota and C_F concentration in food), y_{LB} and y_{LF} are the proportion of lipid in the biota and food, respectively, K_{OW} is the octanol-to-water partition coefficient of the compound involved, and constants a and b are empirical constants for food-to-water and biota-to-water equilibria. Due to the basic similarity of the food-to-water and biota-to-water processes, it would be expected that constants a and b would approach unity. Then

$$BF = y_{LB}/y_{LF}$$

and

$$C_B = (y_{LB}/y_{LF})C_F$$

Also, it can be shown that if C_B and C_F are expressed in terms of lipid content, C_{BL} and C_{FL}, respectively, then C_{BL}/C_{FL} is unity.

This interpretation is supported by the data on the partition processes involved outlined in the previous section. In addition, Forsyth and Peterle[77] have observed a correlation between the concentration of DDT in the stomach and body tissue of some birds. Similarly, Bush et al.[84] have found a significant correlation between toxaphene in the diet and adipose tissue of chickens. This data suggests that the mechanism of biomagnification described in Figure 16 is a major controlling factor in the biomagnification process, and may provide a basis for calculating the biomagnification factor as outlined above.

This model has provided a reasonable explanation for the biomagnification factor in aquatic systems, and with modification, some soil-to-soil invertebrate processes. In these situations, biodegradation was assumed to be negligible. However, Moriarty and Walker[78] point out that terrestrial organisms do not have access to a large volume of water in which to discharge unmetabolized xenobiotic compounds, as do aquatic organisms. This factor has stimulated terrestrial animals to develop metabolic processes to convert lipophilic compounds to hydrophilic compounds, which can be excreted relatively easily. This, illustrated by the data in Figure 19, which indicates that metabolism of biomagnified compounds will be more pronounced with mammals than with fish, may significantly reduce the concentration of compound deposited in body lipid and thus also reduce the Biomagnification Factor.

If biodegradation is significant over the period in which biomagnification occurs, then the observed Biomagnification Factor may be governed more by the rate of biodegradation than the partition process described previously. Walker[79] has developed the following model which is appropriate in these circumstances. If an animal takes up a xenobiotic chemical from food and clears the compound by metabolic processes, then at some stage equilibrium will be reached. The rate of uptake (R_t) can be expressed by the following equation:

$$R_t = (C_F A_f)/N$$

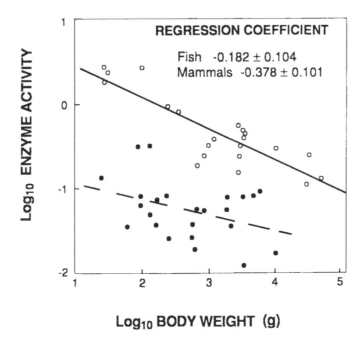

FIGURE 19. Linear regressions for a range of mammalian (○) and fish (●) species enzyme activity and body weight. The activity of the mono-oxygenases associated with liver microsomes is expressed as a proportion of the value for the rat. (From Moriarty, F. and Walker, C. H., *Ecotoxicol. Environ. Saf.*, 13, 208, 1987. Copyright Academic Press. With permission.)

where C_f is the concentration in food, A_f is the fraction of C_F which is absorbed, and N is the number of days required by the organism to consume its own body weight in food. The clearance equation can be expressed in terms of the half life (t_{50}), in days for the initial rate of clearance, and the steady state animal concentration (C_B). If exposure to the chemical ceases then the rate of loss, R_L, which also applies when exposure occurs, can be expressed as

$$R_L = 0.5\, C_B/t_{50}$$

At equilibrium R_t and R_L are in equilibrium so

$$(C_F\, A_f)/N = 0.5\, C_B/t_{50}$$

This can be rearranged to

$$BF = C_B/C_F = (2A_f\, t_s)/N$$

This equation has provided a satisfactory estimate of BF with rats biomagnifying dieldrin from food. The calculated BF and experimentally determined BF were in the range of 0.1 to 0.15. It should be noted that this is generally consistent with the partition method of estimating biomagnification factors when biodegradation occurs. The partition method would suggest a BF of less than unity (on a lipid basis) due to significant biodegradation.

The accumulation of a range of chlorohydrocarbons in the fat of cattle from the diet has been reported by Kenaga.[80] The BF values range from 0.3 to 2.7, indicating a comparatively

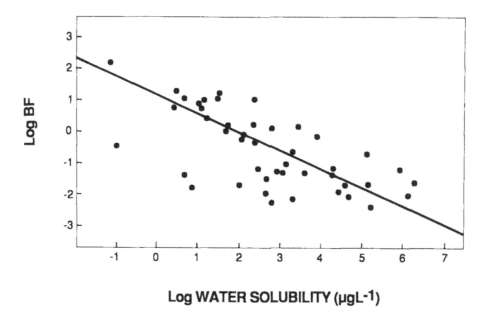

FIGURE 20. Relationship between log BF (C_b in adipose tissue per C_F wet weight) and water solubility ($\mu g\ L^{-1}$) with rats. (From Geyer, H., Kraus, A. G., and Klein, W., *Chemosphere*, 9, 277, 1980. Copyright Pergamon Press. With permission.)

low level of accumulation, consistent with the previously proposed partition mechanism. The differences in values may be due to different rates of biodegradation.

Geyer et al.[81] have evaluated the use of water solubility as a quantitative structure-activity relationship (QSAR) parameter for estimating BF with mammals. The relationship obtained by collating a wide range of data for many different types of compound from the literature is shown in Figure 20. This figure includes many compounds which would be expected to exhibit significant biodegradation, for example, diethylhexyl phthalate, endosulfan, and nitroanisole. If only chlorohydrocarbons are considered, the BF exhibits a limited range from 0.02 to 18.5, i.e., log BF from -1.70 to 1.27. However, a comparatively weak relationship between water solubility (in $\mu g\ L^{-1}$) BF is apparent

$$\log BF = 1.2 - 0.56 \log S \qquad (r = 0.80, \quad N = 40)$$

Geyer et al.[85] have also found BF values (lipid basis) from 3.7 to 24.5 for the biomagnification of 2, 3, 7, 8-TCBD from food by rats, cattle, and monkeys. Somewhat similarly, Kan and Tuinstra[87] found that BF values for biomagnification of some chlorohydrocarbons by hens from their diet ranged from 1.8 to 18.

More recently, Garten and Trabalka[82] have collated literature data on biomagnification by terrestrial organisms, as shown in Table 9. For this analysis the mammals were divided into two classes: first, ruminants, which includes sheep and cows, and second, the nonruminants, including poultry, small birds, rodents, dogs, swine, and primates. The biomagnification factors in all situations are numerically comparatively low, having log K_{ow} values, where the BF value is unity, at 6.91, 7.70, 5.43, and 5.06 for ruminant fat (twice), nonruminant fat, and avian fat. When the BF is unity, then log BF will be zero, and the corresponding low K_{ow} values can be calculated from the equations in Table 9. Since most of the compounds have log K_{ow} values less than these values, BF values will be less than unity.

The correlation coefficients for all of these relationships, including those based on water

TABLE 9
Linear Regression Equations for Biomagnification
Factors of Lipophilic Compounds with Terrestrial Biota

| Variables[a] (log) | | | | |
y	x	Regression equation	Number	r^2
RF	S	$y = -1.476 - 0.495 x$	23	0.67
		$y = 0.191 - 0.608 x$	53	0.54
RF	Kow	$y = -3.457 + 0.500 x$	23	0.62
		$y = -3.935 + 0.511 x$	66	0.34
NRF	Kow	$y = 0.527 - 0.538 x$	37	0.49
NRF	Kow	$y = -3.849 + 0.617 x$	56	0.35
AF	S	$y = 0.990 - 0.451 x$	33	0.48
AF	Kow	$y = -2.743 + 0.524 x$	47	0.54

[a] Where RF is ruminant fat; NPF is non-ruminant fat; AF is avian fat; Kow is the octanol-to-water partition coefficient, and S is the water solubility $(mg \ L^1)$.

From Garten, G. T. and Trabalka, J. R., *Environ. Sci. Technol.*, 17, 590, 1983. Copyright American Chemical Society. With permission.

solubility, are comparatively weak. Travis and Arms[1] calculated beef Biotransfer Factor (B_b) as the ratio between concentration in beef and daily intake of the xenobiotic chemical. A modest correlation between log B_b and log K_{OW} was found for a range of insecticides.

$$\log B_b = 1.033 \log K_{OW} - 7.735 \qquad (r = 0.81, \quad n = 36)$$

Similar to the other data, the B_b values are all comparatively small and cover a limited linear numerical range. It is interesting to note that Villeneuve et al.[86] have reported a dependence on the percentage of transfer of a variety of chlorohydrocarbons from lichen to reindeer as dependent on the log K_{OW} values.

These results are reasonably constant and fit the previously proposed process of partitioning with biodegradation. However, in 1987 Geyer et al.[83] collated literature data on the BF values of lipophilic chemicals in humans. These authors found that the BF values were related to K_{OW} by the following equation:

$$\log BF = 0.0745 \log K_{OW} - 1.19 \qquad (r = 0.969, \quad N = 8)$$

where BF is calculated on a lipid basis.

In addition, an equation based on a parabolic shape was also fitted to the data. Calculation of the log K_{OW} value which corresponds with equal concentration in food and the organisms, in this case, yields a log K_{OW} value of 2.00. Most of the compounds involved have log K_{OW} values greater than this, and thus would be expected to have high values compared to those derived from the equations in Table 9 previously discussed. This result is difficult to explain, but may be related to the method used in calculating BF values.

The previous discussion suggests that there are probably two major factors influencing the biomagnification factor with terrestrial organisms. First, there is the partition mechanism previously described, which would probably result in potential BF values for terrestrial organisms which are of the order of about unity, but more accurately related to lipid contents. Second, there is biodegradation of the biomagnified compounds, which leads to less predictable losses of compounds from the organism and also less predictable BF values. Biodegradation is higher with terrestrial organisms than aquatic organisms, which may explain why the terrestrial BF values are not as consistent with the partition mechanism as the aquatic values. However, a weak dependence on log K_{OW} may occur which could result from differences in the values of constants a and b in the partition processes expressed by Equation 10, or possibly a weak dependence of biodegradation on the K_{OW} values. This indicates that BF values with terrestrial organisms cannot be predicted with accuracy from readily available physicochemical data.

3. Biomagnification in Terrestrial Food Chains

In the previous section, two mechanisms were delineated as probably the most significant in influencing BF values in terrestrial organisms. While one of these, the partition mechanism, can be utilized as a basis for prediction with a reasonable level of confidence, the other, biodegradation, is difficult to predict using currently available methods. In addition, the biodegradation process is likely to differ in different organisms or groups of organisms as a result of different biodegradation capacities. This means that if BF values for one-step terrestrial biomagnification cannot be accurately predicted, then the prediction of BF values in the sequence of steps involved in the food chain cannot be carried out.

The trophic concentration mechanism described by Woodwell[58] involves concentration of persistent xenobiotic chemicals by removal of readily biodegradable food materials, such as fat, within each organism. Repetitions of this at each trophic level results in an increase in concentration with each step in the food chain. Previously, in this chapter, Section III.C.3., it was suggested that in fully aquatic systems, even though biomagnification was occurring, the mechanism controlling the concentrations observed with biomagnification, as well as bioconcentration, was the water-to-organism equilibrium. In terrestrial systems clearly this control is not possible.

A substantial amount of empirical data is available on the concentrations of xenobiotic chemicals in terrestrial systems. Some of this is summarized in Table 10. The mechanism outlined in the previous section suggests that at each step in a terrestrial food chain the maximum concentration obtained by the consumer will be related to the relative lipid content of the food and the consumer. If no biodegradation occurs during the biomagnification process, and if the lipid contents of food and consumer were the same, then the maximum BF value would be unity. Considering that this is a simplification of a complex process, the data in Table 10 is generally consistent with these proposals. Generally, a relationship between trophic status and concentration is not evident, except in the broadest sense. However, there is an exception to this in that birds seem to be consistently high in concentration. In addition, in examining the data in some of the investigations summarized in Table 10 in detail, the relatively high concentrations in birds often provide a major basis for the observed trophic level increases in concentration. Without the bird data the remaining data is much less convincing, and could possibly be explained by differences in such factors as lipid contents, exposure periods, and age of the organisms involved.

Walker et al.[91] have noted the strong biomagnification capacity of fish-eating birds. Birds consuming fish are utilizing a diet which is often consistently high in lipophilic xenobiotic chemicals, as compared to birds consuming terrestrial organisms. Fish are exposed to xenobiotic chemicals through water, and have generally limited capacity to biodegrade these substances. In a number of investigations, Walker et al.[91] have also found that many seabirds have low mono-oxygenase enzyme activity compared to mammals, and so their elimination of lipophilic xenobiotic chemicals is limited. In fact, these authors have suggested this capacity in many seabirds is comparable to that exhibited by fish. This suggests that many seabirds may exhibit a high biomagnification capacity with lipophilic xenobiotic chemicals, as compared to mammals. Moriarty and Walker[78] have extended this discussion to cover a wide range of organisms in that the speed of degradation with biota is the following: small vertebrates > large vertebrates, omnivores/herbivores > predators and mammals >

TABLE 10
Some Concentrations of Xenobiotic Chemicals at Different Trophic Levels in Terrestrial Systems

Compound	Concentrations (mg kg⁻¹ wet weight)				Ref.
Aldrin/dieldrin	Earthworms (1.49)	Insects (0.23—9.67)	Mice 0.98	Snakes (12.35)	2
DDT and metabolites	Marsh plants (0.33—2.80)	Insects (0.23—0.30)	Gulls (3.52—18.5, 75.5)	Cormorant (26.4)	58, 88
PCBs	Fish (0.04—2.1)	Crabs (0.1—0.2)	Gulls (2.6)	Pelican (8.2)	60
DDT (lipid wt)	Herbivorous insects (1—33)	Carnivorous arthropods (1.5—380)	Mice (4.6—5.3)	Shrikes (35—210)	89
γ — BHC	Soil (0.002)	Potatoes (0.01)	Human diet (0.006)	Human fat (0.339)	90
Tetrachlorobenzene, PCB and hexachlorobenzene (dry wt)	Topsoil (0.01—1.05)	Worms (0.03—6.62)	Carnivorous (0.02—3.40)	—	34

birds > fish. These differences can markedly influence the concentration of xenobiotic chemical in different organisms in food chains.

Moriarty and Walker[78] have also pointed out that in field observations of biomagnification in food chains, sometimes increases with trophic level are observed and sometimes not observed. Metabolic activity usually declines with size of organism, as illustrated by data in Figure 19. The observed data have been explained by Moriarty and Walker[78] by the following possibilities:

1. Concentrations increase, but at a declining rate, with each successive member of a food chain.
2. Situations with more rapid elimination of pollutant where intermediate species have the highest concentrations.
3. Situations where the concentration declines with each successive predator.

REFERENCES

1. **Travis, C. C. and Arms, A. D.**, Bioconcentration of organics in beef, milk, and vegetation, *Environ. Sci. Technol.*, 22, 271, 1988.
2. **Korschgen, L. J.**, Soil food-chain pesticide wildlife relationships in aldrin-treated fields, *J. Wildl. Manage.*, 34, 186, 1970.
3. **Butler, J. D., Butterworth, D., Kellow, S. C., and Robinson, H. G.**, Some of the polycyclyic aromatic hydrocarbon (PAH) content of surface soils in urban areas, *Sci. Total Environ.*, 33, 35, 1984.
4. **Nielsen, P. G. and Lokke, H.**, PCDDs, PCDFs and metals in selected Danish soils, *Ecotoxicol. Environ. Saf.*, 14, 147, 1987.
5. **Tinsley, I. J.**, *Chemical Concepts in Pollutant Behaviour*, John Wiley & Sons, New York, 1979, 49.
6. **Spencer, W. F. and Cliath, M. M.**, Volatility of DDT and related compounds, *J. Agric. Food Chem.*, 20, 645, 1972.
7. **Spencer, W. F., Farmer, W. J., and Cliath, M. M.**, Pesticide volatilization, *Res. Rev.*, 49, 1, 1973.
8. **Chiou, C. T. and Shoup, T. D.**, Soil sorption of organic vapors and effects of humidity on sorptive mechanism and capacity, *Environ. Sci. Technol.*, 19, 1196, 1985.
9. **Chiou, C. T., Kile, B. E., and Malcolm, R. L.**, Sorption of vapors of some organic liquids on soil humic acid and its relation to partitioning of organic compounds in soil organic matter, *Environ. Sci. Technol.*, 22, 298, 1988.
10. **Thibodeaux, L. J.**, Chemical exchange rates between air and water, in *Chemodynamics*, John Wiley & Sons, New York, 1979, 300.
11. **Kilzer, L., Scheurnet, I., Geyer, H., Klein, W., and Korte, F.**, Laboratory screening of the volatilization rates of organic chemicals from water and soil, *Chemosphere*, 10, 751, 1979.
12. **Thomas, W.**, Representativity of mosses as biomonitor organisms for the accumulation of environmental chemicals in plants and soils, *Ecotoxicol. Environ. Sci.*, 11, 339, 1986.
13. **Gaggi, C., Bacci, E., Calamari, D., and Fanelli, R.**, Chlorinated hydrocarbons in plant foliage: an indication of the tropospheric contamination level, *Chemosphere*, 14, 1673, 1985.
14. **Buckley, E. A.**, Accumulation of airborne polychlorinated biphenyls in foliage, *Science*, 216, 520, 1982.
15. **Bacci, E., Calamari, D., Gaggi, C., Fanelli, R., Focardi, S., and Morosini, M.**, Chlorinated hydrocarbons in lichen and moss samples from the Antarctic Peninsula, *Chemosphere*, 15, 747, 1986.
16. **Reischl, A., Thoma, H., Reissiger, M., and Hutzinger, O.**, PCDD und PCDF in koniferennadeln, *Naturwissenschaften*, 74, S88, 1987.
17. **Quistad, G. B. and Menn, J. J.**, The disposition of pesticides in higher plants, *Res. Rev.*, 85, 173, 1983.
18. **Thomas, W., Ruhling, A., and Simon, H.**, Accumulation of airborne pollutants (PAH, chlorinated hydrocarbons, heavy metals) in various plant species and humans, *Environ. Pollut.* (Series A), 36, 295, 1984.
19. **Gaggi, C. and Bacci, E.**, Accumulation of chlorinated hydrocarbon vapours in pine needles, *Chemosphere*, 14, 451, 1985.
20 **Connell, D. W. and Hawker, D. W.**, Predicting the distribution of persistent organic chemicals in the environment, *Chem. Aust.*, 53, 428, 1986.

21. **Kenaga, E. E.**, Chlorinated hydrocarbon insecticides in the environment-factors related to bio-concentration of pesticides, in *Environmental Toxicology of Pesticides*, Matsumera, F., Bousch, G. M., and Misato, T., Eds., Academic Press, New York, 1972, 193.

22. **Coulston, F. and Korte, F.**, *Environmental Quality and Safety*, Vol. I. *Global Aspects of Chemistry, Toxicology and Technology as Applied to the Environment*, Georg Thieme, Stuttgart, 1972, 17.

23. **Davis, B. N. K. and French, N. C.**, The accumulation and loss of organochlorine insecticide residues by beetles, worms and slugs in sprayed fields, *Soil Biol. Biochem.*, 1, 45, 1969.

24. **Filov, V. A., Golubev, A. A., Liubiluna, E. I., and Tolokontsev, N. A.**, *Quantitative Toxicology*, John Wiley & Sons, New York, 1979, 94.

25. **Stacey, C. I. and Tatum, T.**, House with organo-pesticides and their levels to human milk — Perth Western Australia, *Bull. Environ. Contam. Toxicol.*, 35, 202, 1985.

26. **Pott, F. and Oberdorster, G.**, Intake and distribution of PAH, in *Environmental Carcinogens: Polycyclic Aromatic Hydrocarbons*, Grimmer, G., Ed., CRC Press, Boca Raton, 1983, 130.

27. **Nash, R. G. and Beall, M. L.**, Chlorinated hydrocarbon insecticides: root uptake versus vapor contamination of soybean foliage, *Science*, 168, 1100, 1970.

28. **Weber, J. B. and Morzek, E.**, Polychloroinated biphenyls phytotoxicity, absorption and translocated by plants and inactivation by activated carbon, *Bull. Environ. Contam. Toxicol.*, 23, 412, 1979.

29. **Strek, H. J. and Weber, J. B.**, Behaviour of polychlorinated biphenyls (PCBs) in soils and plants, *Environ. Pollut.* (Series A), 28, 291, 1982.

30. **Iwata, Y., Gunther, F. A., and Westlake, W.**, Uptake of a PCB (Aroclor 1254) from soil by carrots under field conditions, *Bull. Environ. Contam. Toxicol.*, 11, 523, 1974.

31. **Topp, E., Schenert, I., Attar, A., Korte, A., and Korte, F.**, Factors affecting the uptake of C^{14}-labelled organic chemicals by plants from soil, *Ecotoxicol. Environ. Saf.*, 11, 219, 1986.

32. **Bacci, E. and Gaggi, C.**, Polychlorinated biphenyls in plant foliage: translocation or volatilization from contaminated soils? *Bull. Environ. Contam. Toxicol.*, 35, 673, 1985.

33. **Bacci, E. and Gaggi, C.**, Chlorinated pesticides and plant foliage: translocation experiments, *Bull. Environ. Contam. Toxicol.*, 36, 850, 1986.

34. **Heida, H., Ollie, K., and Prinz, E.**, Selective accumulation of chlorobenzenes, polychlorinated dibenzofurans and 2, 3, 7, 8 — TCBD in wildlife in Volgermeerpolder, Amsterdam, Holland, *Chemosphere*, 15, 1995, 1986.

35. **Wheatley, G. O. and Hardman, J. O.**, Organochlorine insecticide residue in earthworms from ariable soils, *J. Sci. Food Agric.*, 19, 219, 1968.

36. **Haque, A. and Ebing, W.**, Uptake, accumulation and elimination of HCB and 2, 4-D by the terrestrial slug, *Deroceras Merticulatum (Muller)*, *Bull. Environ. Contam. Toxicol.*, 31, 727, 1983.

37. **Davis, B. N. K. and French, N. C.**, The accumulation of organochlorine insecticide residues by beetles, worms and slugs in sprayed fields, *Soil Biol. Biochem.*, 1, 45, 1969.

38. **Gish, C. B. and Hughes, B. L.**, Residues of DDT dieldrin and heptachlor with earthworms during two years following application, *Special Scientific Report — Wildlife No. 241*, Fish and Wildlife Service, United States Department of the Interior, Washington, D.C., 1982.

39. **Lord, K. A., Briggs, G. G., Neal, M. C., and Manlove, R.**, Uptake of pesticides from water and soil by earthworms, *Pest. Sci.*, 11, 401, 1980.

40. **Davis, B. M. K.**, Laboratory studies on the uptake of dieldrin and DDT by earthworms, *Soil Biol. Biochem.*, 3, 221, 1971.

41. **Addison, R. F.**, Organochlorine compounds in aquatic organisms: the distribution, transport and physiological significance, in *Effects of Pollutants on Aquatic Organisms*, Lockwood, B. P. M., Ed., Cambridge University Press, Oxford, 1976, 127.

42. **Grzenda, A. R., Taylor, W. J., and Paris, D. F.**, The elimination and turnover of dieldrin by different goldfish tissues, *Trans. Am. Fish. Soc.*, 101, 686, 1972.

43. **Macek, K. J., Rodgers, J. R., Stolling, D. L., and Corn, S.**, The uptake and elimination of dietary DDT and dieldrin in rainbow trout, *Trans. Am. Fish. Soc.*, 99, 689, 1970.

44. **Chadwick, G. G. and Brocksen, R. W.**, Accumulation of dieldrin by fish and selected fish-food organisms, *J. Wildlife Manage.*, 33, 693, 1969.

45. **Lieb, A. J., Bills, D. B., and Sinnhuber, R. O.**, Accumulation of dietary polychlorinated biphenyls (Arochchlor 1254) by rainbow trout *(Salmo gairdneri)*, *J. Agric. Food Chem.*, 22, 638, 1974.

46. **Bahner, L. H., Wilson, A. J., Sheppard, J. M., Patrick, J. M., Goodman, L. R., and Walsh, G. E.**, Kepone bio-accumulation, loss and transfer through estuarine food chain, *Chesapeake Sci.*, 18, 299, 1977.

47. **Bobroski, C. J. and Epifano, C. E.**, Accumulation of benzo (a) pyrene in a larval bivalve via tropic transfer, *Can. J. Fish. Aquatic Sci.*, 37, 2318, 1980.

48. **Ellegenhausen, H., Guth, J. A., and Esser, H. O.**, Factors determining the bioaccumulation potential of pesticides in the individual compartments of aquatic food chains, *Ecotoxicol. Environ. Saf.*, 4, 134, 1980.

49. **Bruggeman, W. A., Marton, L. B. J. M., Kooiman, D., and Hutzinger, O.,** Accumulation and elimination kinetics of di-, tri- and tetra-chlorobiphenyls by goldfish after dietary exposure, *Chemosphere,* 10, 811, 1981.

50. **Harding, G. C., Vass, W. P., and Drinkwater, K. F.,** Importance of feeding, direct uptake from seawater, and transfer from generation to generation in the accumulation of an organochlorine (DDT) by marine planktonic copepod, *Can. J. Fish. Aquatic Sci.,* 38, 101, 1981.

51. **Skaar, D. R., Johnson, B. P., Jones, J. R., and Huckins, J. N.,** Fate of kepone and mirex in a model aquatic environment sediment, fish and diet, *Can. J. Fish. Aquatic Sci.,* 81, 931, 1981.

52. **Bergman, A., Larsen, G. L., and Bakke, J. E.,** Biliary secretion, retention and excretion of five labelled polychlorinated biphenyls in the rat, *Chemosphere,* 11, 249, 1982.

53. **Muir, B. C. G. and Yarechewski, A. L.,** Dietary accumulation of four chlorinated dioxin congeners by rainbow trout and fathead minnows, *Environ. Toxicol. Chem.,* 7, 227, 1988.

54. **Hilton, J. W., Hodson, P. V., Braun, H. E., Leatherland, J. L., and Slinger, S. J.,** Contaminant accumulation and physiological response in rainbow trout *(Salmo gairdneri)* reared on naturally contaminated diets, *Can. J. Fish. Aquatic Sci.,* 40, 1987, 1983.

55. **Leatherland, J. L. and Sonstegard, R. A.,** Bio-accumulation of organochlorines by yearling coho salmon *(Oncorhynchus kisutch* Walbaum) diets containing great lakes coho salmon and the patho-responses of the recipients, *Comp. Biochem. Physiol.,* 72C, 99, 1982.

56. **Ellgehausen, H., Guth, J. A., and Esser, H. O.,** Factors in the bioaccumulation potential of pesticides in individual compartments of aquatic food chains, *Ecotoxicol. Environ. Saf.,* 4, 134, 1980.

57. **Rudd, R. L.,** *Pesticides and the Living Landscape,* University of Wisconsin Press, Madison, 1964.

58. **Woodwell, G. M.,** Toxic substances and ecological cycles, *Sci. Am.,* 216, 24, 1967.

59. **Bender, M. E. and Huggett, R. J.,** Fate and effects of kepone in the James River, in *Reviews in Environmental Toxicology I,* Hodgson, E., Ed., Elsevier Science, Amsterdam, 1984, 5.

60. **Shaw, G. R. and Connell, D. W.,** Factors influencing concentrations of polychlorinated biphenyls in organisms from an estuarine ecosystem, *Aust. J. Mar. Freshwater Res.,* 33, 1057, 1982.

61. **Gunkel, G. and Streit, B.,** Mechanisms of bioaccumulation of a herbicide (Atrozine, s-trizane) in a freshwater mollusc *(Ancylus fluviatilis Mull)* and a fish *(Coregonus fera* Jurine), *Water Res.,* 14, 1573, 1980.

62. **Eberhardt, L. L., Meeks, R. L., and Peterle, T. J.,** Food chain model for DDT kinetics in a freshwater marsh, *Nature,* 230, 60, 1971.

63. **Isaacs, J. D.,** Potential trophic biomasses and trace-concentrations in unstructured marine food webs, *Mar. Biol.,* 22, 97, 1973.

64. **Norstrom, R. J., McKinnon, E. A. E., and DeFreitas, A. S. W.,** A bioenergetics-based model for pollutant accumulation by fish. Stimulation of PCB and methylomercury residue levels in Ottawa River yellow perch *(Perca flavescens),* J. Fish. Res. Board Can., 33, 248, 1976.

65. **Fagerstrom, T.,** Body weight, metabolic rate, and trace substances turnover in animals, *Oecologia (Berlin),* 29, 99, 1977.

66. **Thomann, R. V.,** Equilibrium model of fate of micro-organisms in diverse aquatic food chains, *Can. J. Fish. Aquatic Sci.,* 38, 280, 1981.

67. **Jensen, A. L., Spigarelli, S. A., and Thommes, M. N.,** PCB uptake by five species of fish in Lake Michigan, Green Bay of Lake Michigan, and Cayuga Lake, New York, *Can. J. Fish. Aquatic Sci.,* 39, 700, 1982.

68. **Thomann, R. B. and Connolly, J. P.,** Model of PCB in the Lake Michigan lake trout food chain, *Environ. Sci. Technol.,* 18, 65, 1984.

69. **Connolly, J. P. and Pedersen, C. J.,** A thermo-dynamic evaluation of organic chemical accumulation in aquatic organisms, *Environ. Sci. Technol.,* 22, 99, 1988.

70. **Henny, C. J. and Meeker, B. L.,** An evaluation of blood plasma for monitoring DDE in birds of prey, *Environ. Pollut.* (Series A), 25, 291, 1981.

71. **Wolff, M. S., Thornton, J., Fischbein, A., Lilis, R., and Selikoff, I. J.,** Disposition of polychlorinated biphenyl congeners in occupation exposed persons, *Toxicol. Appl. Pharmacol.,* 62, 294, 1982.

72. **Brown, J. F. and Lawton, R. W.,** Polychlorinated biphenyl (PCB) partitioning between adipose tissue and serum, *Bull. Environ. Contam. Toxicol.,* 33, 277, 1984.

73. **Kan, C. A. and Rooyen, J. C. J.,** Accumulation and depletion of some organochlorine pesticides in broiler breeder hens during the second laying cycle, *J. Agric. Food Chem.,* 26, 465, 1978.

74. **Tanabe, S., Subramanian, A. N., Hidaka, H., and Tatsukawa, R.,** Transfer rates and pattern of PCB isomers and congeners and DDE from mother to egg in Adelie penguin *(Pygoscelis adeliae), Chemosphere,* 15, 343, 1986.

75. **Norstrom, R. J., Clarke, T. P., Jeffrey, D. A., and Won, H. T.,** Dynamics of organochlorine compounds in herring gulls *(Larus argentatus):* I. Distribution and clearance of DDE in free-living herring gulls *(Larus argentatus), Environ. Toxicol. Chem.,* 5, 41, 1986.

76. **Kan, C. and Rooyen, J. C. J.**, Accumulation and depletion of some organochlorine pesticides in high-laying hens, *J. Agric. Food Chem.*, 26, 935, 1978.

77. **Forsyth, B. J. and Peterle, T. J.**, Species and age differences in accumulation of DDT by voles and shrews in the field, *Environ. Pollut.* (Series A), 33, 327, 1984.

78. **Moriarty, F. and Walker, C. H.**, Bioaccumulation in food chains — a rational approach, *Ecotoxicol. Environ. Saf.*, 13, 208, 1987.

79. **Walker, C. H.**, Kinetic models for predicting bioaccumulation of pollutants in ecosystems, *Environ. Pollut.*, 44, 227, 1987.

80. **Kenaga, E. E.**, Correlation of bioconcentration factors of chemicals in aquatic and terrestrial organisms and their physical and chemical properties, *Environ. Sci. Technol.*, 14, 553, 1980.

81. **Geyer, H., Kraus, A. G., and Klein, W.**, Relationship between water solubility and bioaccumulation potential of organic chemical in rats, *Chemosphere*, 9, 277, 1980.

82. **Garten, C. T. and Trabalka, J. R.**, Evaluation of models for predicting terrestrial food chain behaviour xenobiotic, *Environ. Sci. Technol.*, 17, 590, 1983.

83. **Geyer, H. J., Scheunert, I., and Korte, F.**, Correlation between the bioconcentration potential of organic environmental chemicals in humans and their n-octanol/water partition coefficient, *Chemosphere*, 16, 239, 1987.

84. **Bush, P. B., Tanner, M., Kiker, J. T., Page, R. K., Booth, N. H., and Fletcher, O. J.**, Tissue residue studies on toxaphene in broiler chickens, *J. Agric. Food Chem.*, 26, 126, 1978.

85. **Geyer, H. J., Scheunert, I., Filser, J. G., and Korte, F.**, Bioconcentration potential (BCP) of 2, 3, 7, 8-tetrachlorodibenzo-p-dioxin (2, 3, 7, 8-TCDD) in terrestrial organisms including humans, *Chemosphere*, 15, 1495, 1986.

86. **Villeneuve, J. P., Holm, E., and Cattini, C.**, Transfer of chlorinated hydrocarbons in the food chain: lichen, reindeer, man, *Chemosphere*, 14, 1651, 1985.

87. **Kan, C. A. and Tuinstra, L. G. M. T.**, Accumulation and excretion of certain organochlorine insecticides in broiler breeder hens, *J. Agric. Food Chem.*, 24, 775, 1976.

88. **Woodwell, G. M., Wurster, C. F., and Isaacson, P. A.**, DDT residues in an east coast estuary — a case of biological concentration of a persistent insecticide, *Science*, 156, 821, 1967.

89. **Rudd, R. L., Craig, R. B., and Williams, W. F.**, Trophic accumulation of DDT in a terrestrial food web, *Environ. Pollut.* (Series A), 25, 219, 1981.

90. **Szokolay, A., Rosival, L., Uhnak, J., and Madaric, A.**, Dynamics of benzene hexachloride (BHC) isomers and other chlorinated pesticides in the food chain and in human fat, *Ecotoxicol. Environ. Saf.*, 1, 349, 1977.

91. **Walker, C. H., Knight, G. C., Chipman, J. K., and Ronis, M. J. J.**, Hepatic microsomal monooxygenases of sea birds, *Mar. Environ. Res.*, 14, 416, 1984.

92. **Jarvinen, A. W., Hoffman, M. J., and Thorslund, T. W.**, Long-term toxic effects of DDT food and water exposure on fathead minnows *(Pimpales promelas)*, *J. Fish. Res. Board Can.*, 34, 2089, 1977.

Chapter 8

BIOACCUMULATION OF METALLIC SUBSTANCES AND ORGANOMETALLIC COMPOUNDS

Darryl W. Hawker

TABLE OF CONTENTS

I. INTRODUCTION

Metals in various forms are natural constituents of water bodies, both marine and freshwater, and are derived from natural processes such as the erosion of ore-bearing rocks and volcanic activity. Anthropogenic activities constitute an increasingly important source of metals in the aquatic environment. In recent years, the impact of increased levels of metals on biota has been the focus of much research. Metals are accumulated by all aquatic organisms to varying degrees. In some cases, biotic metal concentrations can reach lethal levels, while in others sublethal levels may still represent a problem in terms of biomagnification. The Minamata Bay incident, in which 43 people died from eating mercury contaminated seafood, serves to highlight the importance of, and potentially serious consequences associated with, bioaccumulation of metals. In order to understand this process, it is necessary to investigate the chemical, physiological, and environmental factors which control the accumulation of metals by organisms.[1]

Many metals, in trace amounts, are essential for living organisms. As examples, the respiratory pigments of vertebrates and many invertebrates contain iron, molluscs and higher crustaceans contain copper, and tunicates contain vanadium. In addition, many enzymes contain metals such as zinc and cobalt.[2]

The rocks and soils directly exposed to surface waters are usually the largest natural source of metals in the aquatic environment. Precipitation and atmospheric fallout, which remove particulate matter or aerosols from the atmosphere, are other important sources. Dead and decomposing vegetation, together with animal matter, also contribute small amounts of metals to adjacent waters.[3] Anthropogenic inputs include domestic sewage, mining and mine drainage effluent, and industrial discharges.

Aqueous metal concentrations vary considerably throughout the world, depending on factors such as proximity to major input sources, temperature, and salinity. Mean background levels in uncontaminated seawater and freshwater are presented in Table 1. The concentrations of sodium, magnesium, potassium, and calcium are much greater in seawater, but the concentrations for the metals in Table 1 are within an order of magnitude of each other except for iron and aluminum.[4]

Metals in aquatic systems can precipitate if the concentration of a metal in ionic form is higher than the solubility of the least soluble compound that can be formed between the metal cation and anions present in the water, such as carbonate, hydroxyl, or chloride. In addition, metals can adsorb to particulates entrained in the water column, or sediments. Adsorption of metals from solution often occurs at the surfaces of materials such as clay, hydrated ferric oxide, and hydrated manganese dioxide.[5] Compared with overlying water, sediments may contain very high concentrations of metals. Since different metals are not equally readily precipitated or adsorbed, some will tend to be deposited near the input source, whereas other more soluble ones may be dispersed over a wider area. Eventually, metals contained in sediment will be recycled into the overlying water, initially by dissolution into the interstitial water phase.[6]

Metals are also removed from solution through bioconcentration by aquatic organisms. Bioconcentration factors (K_B) for different metals vary considerably from species to species. Table 2 contains some typical metal bioconcentration factors for selected marine and freshwater organisms.[3] As shown, K_B values of the order of 10^2 and 10^3 are commonly found, with some 10^5 or more.

Other recognized processes for accumulation of metals in biota include ingestion of suspended particulate material containing adsorbed metals from the surrounding water and ingestion of food material containing metals.[7] In addition, since many organisms live and feed in sediments, they may absorb metals, either directly from the sediment, or via the interstitial water.[5] Therefore, it should be remembered that there are other sources, apart from aqueous solution, from which aquatic biota may accumulate metallic substances.

TABLE 1
Typical Background
Concentrations of Some Metals in
Rivers and Oceans

Metal	River (ppb)	Ocean (ppb)
Ag	0.3	0.1
Al	400	5
As	1	2.3
Cd	0.03	0.05
Co	0.2	0.02
Cr	1	0.6
Cu	5	3
Fe	670	3
Hg	0.07	0.05
Mn	5	2
Mo	1	10
Ni	0.3	2
Pb	3	0.03
Sb	1	0.2
Se	0.2	0.45
Sn	0.04	0.01
V	1	1.5
Zn	10	5

From Bryan, G. W., in *Marine Pollution*, Johnston, R. B., Ed., 1976, Chap. 3. Copyright Academic Press. With permission.

TABLE 2
Typical Metal Bioconcentration Factors of Selected Aquatic Organisms

Metal	Marine organisms					Freshwater organisms		
	Phytoplankton	Zooplankton	Macrophytes	Molluscs	Fish	Macrophytes	Molluscs	Fish
Ti	2,700	—	—	—	—	—	—	—
Cr	7,800	—	2,880	21,800	—	—	267	10
Mn	3,800	3,900	—	2,300	373	1,450	—	23
Fe	28,300	114,600	—	14,400	—	3,642	—	190
Ni	570	560	1,050	4,000	235	—	650	85
Co	—	—	—	—	50	1,367	300	90
Cu	2,800	1,800	2,890	3,800	127	158	1,500	60
Zn	5,500	8,800	7,000	27,300	533	318	2,258	228

From Williams, S. L., Aulenbach, D. B., and Clesceri, N. L., in *Aqueous Environmental Chemistry of Metals*, Rubin, A. J., Ed., 1976, Chap. 2. Copyright Ann Arbor Science Publishers. With permission.

II. CHEMICAL FORMS OF METALS

The chemical form in which a metal is found in water has an important bearing on its availability to an aquatic organism. Organometallic species, for example, have often been shown to be the predominant form bioaccumulated for metals where such compounds are relatively stable.[8,9] In general, native metals are insoluble in water, and they exist in the environment as ions.

Mercury dissolved in seawater is in the form of the mercuric (Hg^{2+}) ion, and occurs

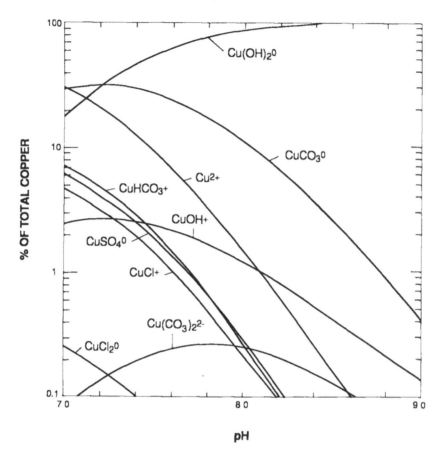

FIGURE 1.   Calculated distribution of species of copper in seawater at 25°C and one atmosphere, as a function of pH. (From Zirino, A. and Yamamoto, S., *Limnol. Oceanogr.*, 17, 661, 1972. Copyright American Society of Limnology and Oceanography. With permission.)

largely as $Hg(OH)_2$ and $HgCl_2$. Bryan[6] also considers that complex halide ions such as $[HgCl_4]^{2-}$ exist in solution. Mercury forms stable complexes with organic compounds occurring in water, especially sulphur containing proteins and humic substances. To a considerable extent, however, mercury is absorbed onto particulate matter, and under anaerobic conditions in the sediment may be present as HgS and HgS_2.[2]

Cadmium is strongly associated with chloride ions in seawater, and the predominant species have been calculated to be $CdCl_2$ (51%), $[CdCl]^+$ (39%), $[Cd Cl_3]^-$ (6.1%) and Cd^{2+} (2.5%). In marine sediments, it is likely that this species distribution will exist also.[10]

Zirino and Yamamoto[10] found that pH had a considerable influence on copper speciation in seawater. This is illustrated in Figure 1, which shows the percentage of total copper plotted against pH. At the average marine pH of 8.1, the chemical species of copper existing in solution are $Cu(OH)_2$ (90%) and $Cu CO_3$ (8%), with the uncomplexed cupric ion (Cu^{2+}) and $[CuOH]^+$ each representing about 1%.[10] The speciation of copper in freshwater is dependent on a number of factors, particularly alkalinity, pH, and hardness, but in general, the most abundant forms of soluble copper are $CuCO_3$, Cu^{2+}, and $Cu(OH)_2$.[11] Copper is one of the metals most readily removed from solution by absorption to particulates, and it is estimated that 83% of the copper in the marine environment is sorbed to some material.[2] Since marine sediments are lower in pH than the overlying water, on the basis of Figure 1 it would be expected that species such as $[CuHCO_3]^+$ and $CuSO_4$ would become more important in sediments.[10] Other metals, such as lead and zinc, show similar compositional variability in solution and adsorbed onto sediments. This, in turn, may affect the relative dominance of the various bioaccumulation pathways.

III. PATHWAYS AND MECHANISMS OF BIOACCUMULATION

A. BIOCONCENTRATION

Metals can be bioconcentrated by aquatic organisms either via active or passive mechanisms. Active mechanisms have been demonstrated for sodium, potassium, calcium, and magnesium. Over long periods of evolution, ever present components of the aquatic environment have become incorporated into vital physiological functions. Examples are respiratory pigment components (iron and copper) and metalloenzyme constituents (zinc).[7] It is likely that these metals have evolved active transport mechanisms as well.

For other metals, uptake from solution is generally considered to occur via active transport systems in algae and phytoplankton.[2] In these cases, ion-exchange processes involving proteins are usually involved,[6,12] although it recently has been postulated that nickel coordinates to functional groups in extracellular mucopolysaccharides with some strains of algae.[13] In filter feeders, such as bivalve molluscs and tunicates, the mucous sheets of the ciliary feeding mechanisms have been implicated in bioconcentration. Passive diffusion may occur across concentration gradients created by adsorption on the mucus.[2,6]

With many higher organisms, the nature of the metal transport system is unclear. It has been suggested that if a passive mechanism is operating, then the observed bioconcentration factors should parallel the stabilities of complexes formed between the metals and organic ligands (the so-called Irving-Williams series):[6,12,14]

$$Pd > Cu > Ni > Pb > Co > Zn > Fe > Cd > Hg$$

While in some cases, this order is observed, in many others it is not.

It would not be surprising to find active, specific transport mechanisms for metals belonging to the same Periodic Table group as those such as potassium and calcium, where active transport almost certainly occurs. In support of this, ^{90}Sr is thought to bioconcentrate in the same manner as calcium, and ^{137}Cs is probably transported by the potassium pump in the isopod *Sphaeroma hookeri*.

There is increasing evidence to suggest that cadmium may be taken up by an active process, also. Experiments with the mussel *Mytilus edulis* have shown an initial lag period. Prior complex formation of the cadmium with EDTA, pectin, or humic and alginic acids increases both the rate of uptake and final tissue concentrations, and eliminates the lag time. These effects are consistent with the theory that the metal is not transported across membranes as the ionic species, but rather complexed by a ligand, possibly thionein. The initial lag period then represents the exchange of cadmium from ionic to complexed forms, which must be rate-limiting.[15]

Active transport was also considered to be the reason why an apparent equilibrium developed between aqueous and tissue cadmium concentrations in bioconcentration experiments involving bluegill and largemouth bass. The cadmium was suggested to inhibit enzymes controlling the carrier system of active transport, leading to cessation of metal accumulation and an effective equilibrium.[16] Further evidence is provided by the absence of bioconcentration of cadmium, after rendering freshwater amphipods *(Gammarus pulex)* moribund by treatment with 2,4-dinitrophenol. This, together with the observation of a negative relationship between cadmium uptake rate and the calcium concentration of the animal, suggests that cadmium bioconcentration may occur by cadmium substituting for calcium in a calcium regulatory mechanism.[17] Further investigations need to be undertaken, however, before it is apparent which transport mechanism predominates in bioconcentration of metals.

B. BIOACCUMULATION FROM SUSPENDED PARTICULATES

Metals in natural water exist either in suspended or colloidal forms or in solution. Williams et al.[3] suggest that suspended particles are those greater than 0.1 μm in diameter, while Giesy et al.[18] consider 0.15 μm to be the boundary between colloids and particulates. The form in which metals occur in water influences their accumulation by aquatic organisms, and some metals exist largely in a particulate form. Iron in seawater, for example, is predominantly particulate (as $Fe_2O_3 \cdot H_2O$), while manganese is also largely particulate.[19]

Accumulation of metals such as cadmium, copper, lead, chromium, and zinc by the Pacific oyster *(Crassostrea gigas)* has been found to primarily involve bioaccumulation of particulates, rather than bioconcentration. In bivalve molluscs of the class Lamellibranchia, it is known that particulates pass through the gills and are trapped by secretions of the hypobranchial glands.[20] Investigations concerning the mussel *Mytilus edulis* have also shown the importance of bioaccumulation of suspended particulates, whether through the digestive system or uptake through the gills by pinocytosis.[21]

C. BIOACCUMULATION FROM FOOD

There is some uncertainty regarding the relative importance of food as a source of metals for aquatic animals.[7] Uptake from food particles by the filter feeding *Daphnia magna* has been found to contribute substantially to the total zinc body burden, while the metal content of corals is thought to be related in part to the feeding characteristics of the coral polyps.[22,23] With bivalves, it is generally assumed that food and particulates are more important sources that direct bioconcentration from water. This is also the case with many gastropods.[7] Recent investigations involving crustaceans, however, have suggested that uptake via food may not be as significant as first was thought. In addition, several authors have found that cadmium and lead are mainly bioconcentrated by several fish species.[22,24,25]

An explanation for the variable contribution of food to biotic metal levels may be found in the variable feeding rate of consumers, and the different kinds of food ingested. Other factors, such as the efficiency with which the food material is digested and the metal absorbed, the chemical form of the metal, and the binding site in food would all be expected to play a role.[6] With cadmium, for example, it is generally found that bioaccumulation from solid matter such as food is much less efficient than from the aqueous phase.[26] Nimmo et al.[27] calculated that some 1.5×10^4 times more cadmium must be introduced in food compared to water to produce the same effect in shrimps. The importance of accumulation from diet may also depend on aqueous metal concentration, since at lower aqueous concentrations a relatively higher amount may be bioaccumulated.

Overall, on the basis of numerous field studies, it is evident that the concentration of metals (with the notable exception of mercury) mostly decreases in consumers of higher trophic levels, and thus biomagnification of metals generally does not occur in aquatic food chains.[6,22]

D. BIOACCUMULATION FROM SEDIMENTS AND INTERSTITIAL WATER

The behavior of metals in sediments has been the focus of increasing attention in recent years. Metals bound to sedimentary material are probably in a state of dynamic equilibrium with metals in overlying and interstitial water. If the interstitial water is rich in organics capable of forming soluble complexes with metals, interstitial aqueous metal levels may be many times that of the overlying water. The ratio of sediment to overlying water concentration can be as high as 5×10^4, though this depends on the type of sediment, its state of subdivision, period of contact, temperature, and pH, as well as salinity and hardness of water.[7]

Contaminated sediments may persist as sources of metals along after the original source has been removed, through remobilization and upward diffusion. In addition to storage, chemical transformations may occur in sediment. Precipitated ferric oxide and manganese dioxide are often reduced in the deeper anaerobic portions of sediment to the manganous and ferrous forms which are much more soluble. Another important example is mercury. Originally released into the environment as divalent or metallic mercury, oxidation, then bacterial biomethylation, occurs in the upper layers of sediment, forming the more toxic mono- and dimethyl mercury species.[6] Other metals which undergo biomethylation in sediment include tin, palladium, platinum, gold, and thallium.[7]

Large bioconcentration factors for metals with oysters of the order of 1×10^5 which are found in the literature have been criticized because bioaccumulation from sediments, or bioconcentration from interstitial water, may be more important processes.[14,20] Bioaccumulation factors based on sediment are often considerably lower due to the higher metal content of the sediment. Few reports of K_B values associated with interstitial water exist due to practical problems associated with their measurement, but it would be expected that they would also be smaller in magnitude than K_B values derived on the basis of overlying water.

For benthic organisms, it is often difficult to determine the relative contributions of accumulation from ingested sediment or interstitial water toward biotic metal levels. To investigate sedimentary bioaccumulation, Tessier[28] sequentially fractionated sorbed metals from Canadian lakes into (1) exchangeable metals, (2) metals bound to carbonates, (3) metals bound to iron-manganese oxides, (4) metals bound to organic matter and sulfides, and (5) residual metals. Metal levels in various tissues (or the whole organism) of the benthic pelecypod *Elliptio complanata* were found to be directly related not to total sediment metal concentrations, but to levels in one or more of the individual fractions. Correlations were better for relatively easily extractible metals, suggesting that bioavailability is inversely related to metal binding strength to the various sediment fractions.[28] This highlights the influence of sediment composition on the relative importance of sedimentary bioaccumulation.

In an experiment to determine interspecial variability in accumulation pathways, three burrowing coastal benthic species, a polychaete (*Aremicola marina*), a bivalve mollusc (*Scrobicularia plana*), and a crustacean (*Corophium volutator*) were exposed to [241]Am (III) and [238]Pu (VI) contaminated sediment and seawater, in separate experiments. The *C. volutator* consistently accumulated metals to the greatest extent from both sediment and seawater. Data obtained suggests that all test organisms bioaccumulate from ingested sediment to the same degree, and that the difference in observed biotic metal concentrations is due to different abilities to bioconcentrate Americium and Plutonium from interstitial water and seawater.[29]

It is probable that benthic organisms bioaccumulate from both sediment and interstitial water phases, in most cases. General statements concerning the relative importance of the two pathways are difficult, since it is dependent on a number of biological, chemical, and physical factors.

E. STORAGE AND ELIMINATION

There is great variability in the tolerance of aquatic organisms toward metals. Some, such as molluscs, bioaccumulate metals to levels considerably higher than in the ambient water and sediments; levels that would be toxic to other organisms. Bivalve molluscs, e.g., *Mytilus edulis*, have been widely advocated and adopted for extended monitoring of metal levels in the aquatic environment. Programs such as "Mussel Watch" are good examples of this.[30] Additionally, bioaccumulated metals are often eliminated very slowly.

In considering the storage of metals, it is important to realize that they are not stored uniformly within organisms. After exposure to cadmium, for example, the marine crustacean *Lysmata seticaudata* showed the following order of cadmium concentrations in the body: viscera > exoskeleton > muscle > eyes. Under similar conditions, organs and tissues of the fish *Fundulus heteroclitus* exhibited an order of liver and kidney > gills > head > remainder.[31] Studies into the nature of zinc, copper, and cadmium present in the liver and kidneys of terrestrial mammals have shown that these metals are stored bound to a relatively low molecular weight, high cysteine content protein called metallothionein. Subsequent investigations have revealed the presence of similar proteins in algae, molluscs, fish, and marine mammals. In fact, it has been suggested that metallothionein-like proteins are ubiquitous to the living world.[31,32] Metallothionein levels also seem to be inducible, which means that initial exposure to low concentrations of a metal may induce protection against subsequent exposure to higher concentrations.[33] The strong binding of some metals such as cadmium to metallothioneins may play a role in their long biological half-lives.

Other methods of storage include deposition in specific organs and skeletal material, such as bone, as well as intracellular deposition. The green flagellate *Dunaliella tertiolecta* is very tolerant towards mercury since it produces H_2S, and the metal is precipitated within the cell as the relatively insoluble sulfide. Extremely high concentration of zinc, copper, and manganese occurs in the renal organs of molluscs, in the form of granules. In many cases, copper is associated with sulfur, suggesting a combination with a protein or sulfide, while zinc occurs as zinc phosphate.[5,21] In oysters, copper is stored as granules in leukocytes, and it seems likely that other metals may be stored in the same way, then gradually eliminated.

One elimination mechanism is back across the body surface or gills, such as occurs in the shore crab *(Carcinus maenus)* and the rainbow trout *(Salmo gairdnerii)*. For some metals such as mercury, elimination is assisted by metabolic processes. In many fish, most of the mercury is in a methylated form, complexed by metallothionein. In tissues such as the liver, however, slow demethylation to divalent mercury (Hg^{2+}) occurs.[34] This substance is more water soluble, and elimination takes place via the urine. Fish eliminate cadmium through the kidneys, and metal accumulation has been linked with renal failure in some cases.[33] Crustaceans and molluscs are also able to excrete metals such as zinc, copper, cobalt, manganese, and mercury in the urine. Another method is by excretion of metal into the gut. This is the principal elimination route for a number of molluscs and crustaceans, such as the freshwater crayfish and the barnacle. Finally, for oysters, leukocytes have the ability to transport metals to the surface epithelium tissues from where elimination may occur.[7]

F. MODELS FOR METAL ACCUMULATION

The metal levels found in aquatic organisms depend on a complex accumulation history that is influenced by external (exogenous) as well as internal (endogenous) factors. Models attempting to describe metal bioaccumulation are often more complex than those describing the behavior of hydrophobic organic chemicals.

As an example, the bioaccumulation of mercury by fish has been modeled assuming a single residue pool, and that uptake and clearance are functions of metabolic rate. In addition, three endogenous variables, body weight, growth rate, and body burden are employed. The accumulation rate is equal to the difference in gain and loss fluxes. If mercury enters a fish, both through contaminated food and by direct uptake from solution, the gain flux is divisible into dietary and aqueous terms, both of which are functions of metabolic rate.

The metabolic rate of animals (M) can be related to body weight (W) by

$$M = a\ W^b$$

where a and b are constants. The dietary term may be expressed as a sum of two terms, one proportional to metabolic rate, and the other to the rate of change of body weight reflecting food used for metabolic and growth purposes, respectively. Bioconcentration via the gills is assumed to be proportional to oxygen consumption, which is related to body weight. Therefore uptake may be expressed as

$$p_1 W^b + p_2 \frac{dW}{du}$$

where p_1 and p_2 represent coefficients relating uptake of mercury to W and dW/du, and u is the age of the fish.

Loss of mercury is assumed to be first order, with the associated rate constant varying directly with metabolic rate. Overall then, the rate of change of mercury levels with age is

$$\frac{dC}{du} = p_1 W^b + p_2 \frac{dW}{du} - p_3 W^{-(1-b)} C$$

where p_3 is a clearance parameter relating the elimination of mercury to body weight and body burden.[35] Comparison of mercury concentrations calculated using this model with experimental data has revealed reasonable correspondence.

The bioconcentration of metals such as cadmium, copper, lead, and mercury by mussels has been interpreted based on the following equilibrium between the metal in solution, and specific binding sites within the mussel tissue:

$$C_W + B \underset{k_2}{\overset{k_1}{\rightleftharpoons}} CB$$

In this expression C_W is the aqueous metal concentration, B is the concentration of the free metal binding sites within the organism, CB is the concentration of binding sites occupied by the metal, and k_1 and k_2 are the uptake and clearance rate constants, respectively. The total concentration of metal binding sites (B_o) is therefore $B + CB$. At equilibrium, $dCB/dt = 0$ and the equilibrium constant for the process is given by

$$K = k_1/k_2 = \frac{CB}{C_W \cdot B}$$

Substitution of $B_o = B + CB$ into this equation and rearrangement yields

$$CB = \frac{B_0}{1 + 1/KC_W} \qquad (1)$$

The bioconcentration factor (K_B) is by definition CB/C_W, and so from Equation 1

$$K_B = \frac{B_0}{C_W + 1/K}$$

$$\text{or} \qquad \frac{1}{K_B} = \frac{1}{KB_0} + \frac{C_W}{B_0}$$

This three parameter (k_1, k_2, and B_o) model thus predicts a direct linear relationship with slope $1/B_o$ and intercept $1/KB_o$ between $1/K_B$ and C_w, and to a large extent was found to account for relevant experimental results.[36]

Ruzic[37] has suggested that the bioaccumulation of various radionuclides such as ^{137}Cs and ^{51}Cr by molluscs and crustaceans can generally be explained using a two compartment organism model, in which each compartment is in dynamic equilibrium with the ambient water and the other internal compartment.

A number of strictly empirical models have also been developed in an attempt to correlate the bioaccumulation behavior of metals by aquatic organisms. For example, field studies of metal bioaccumulation by mussels (*Mytilus edulis*) have enabled the construction of a statistical model relating trace metal levels of copper and zinc in seawater and sorbed onto particulate matter to metal concentrations in mussels. Variables considered, apart from copper and zinc concentration in the mussel, are concentrations of iron, strontium, and lead, together with water content in the mussel, dry weight of the mussel tissues, and time of collection (Day').[38]

The logarithm of copper concentration as suspended particulate material can be expressed as

$$-0.453 - 0.2171 \log [Cu] - 0.3076 [Day'] + 1.5212 \log [Fe]$$
$$+ 0.4566 \log [Weight] - 0.5682 \log [Sr] - 0.198 \log [1 + Pb]$$

while the logarithm of aqueous copper concentration is given by

$$1.9218 - 0.8192 \log [Cu] + 0.1012 [Day'] - 0.5334 \log [Fe] - 0.3691 \log [Zn]$$

Approximately 50% of the variance of the soluble metal levels, and somewhat less of the variance of the suspended particulate matter sorbed metal levels, can be accounted for by this model. It might be expected that the model could be improved by inclusion of factors such as salinity and turbidity. However, the relatively low predictive ability highlights the fact that, in general, the bioaccumulation of metals is not as amenable to modeling as the bioaccumulation of hydrophobic organic xenobiotics.

IV. THE INFLUENCE OF ENVIRONMENTAL FACTORS ON BIOACCUMULATION

A. TEMPERATURE

Any dependence of bioaccumulation on temperature could have a significant impact on observed biotic levels of metals, particularly in temperate areas where organisms are exposed to widely varying seasonal temperatures. Numerous investigations have shown that, in general, accumulated metal concentrations increase with temperature. For example, Jackim et al.[39] found that an increase in the ambient water temperature from 10°C to 20°C resulted in a measurable increase in cadmium levels with marine bivalves, while Rodgers[40] observed increasing mercury accumulation by rainbow trout with temperature. Some authors have concluded that the effects of temperature are based on a change in the organism's metabolic rate; however, this cannot be a general effect because in some situations, such as the uptake of ^{109}Cd by the mussel *Mytilus edulis*, and ^{75}Se by shrimp, no temperature dependence is seen. The effect of temperature variance may be exerted in a more subtle way, perhaps involving ion transport mechanisms at membrane surfaces.[34,39,40] Other environmental factors can be important, as Phillips[41] found, for instance, that the bioaccumulation of zinc, cadmium, and copper by mussels was temperature-dependent only in conjunction with relatively hyposaline ambient water.

With organisms such as algae, an additional complicating factor may be present. Increasing temperatures generally promote faster growth. Should such conditions also promote metal accumulation, one effect may tend to balance the other, so that no temperature influence on bioaccumulation is observed. In some cases, growth may be so rapid that metal levels actually fall with temperature due to a growth dilution effect.[34]

B. SALINITY

Studies of the effect of salinity on the bioaccumulation of metals have shown that in

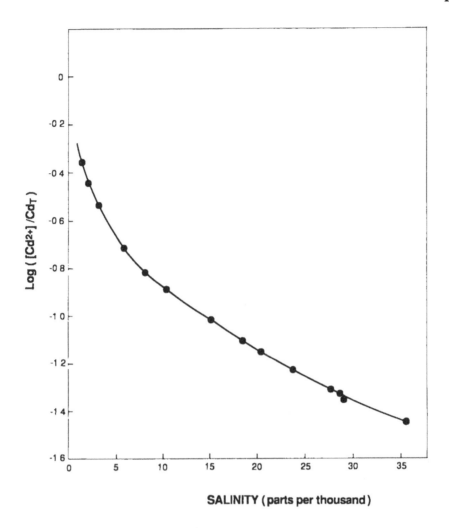

FIGURE 2. The logarithm of the fraction of total dissolved cadmium present as the free cadmium ion (Cd^{2+}), as a function of salinity. (From Sunda, W. G. et al., *Environ. Sci. Technol.*, 12, 409, 1978. Copyright American Chemical Society. With permission.)

general, biotic metal concentrations are greater with decreasing salinity. A salinity decrease from 30 parts per thousand to 20 parts per thousand, for example, produced a 400% increase in cadmium bioaccumulated by marine bivalves.[39] Experiments with perfused rainbow trout gills have shown that substances such as $CdCl^+$ and $Cd\ Cl_2$ are not readily taken up;[42] however, the proportion of these forms relative to total cadmium decreases with salinity (Figure 2). An inverse relationship between bioaccumulated cadmium and salinity is also consistent with observations that organisms collected along a salinity gradient in an estuary have decreased levels of cadmium as the water becomes more saline.[1]

It has been suggested that stable ambient salinities may influence the metal content of an aquatic organism in two ways. First, many metals are rendered more available in areas of low salinity because of the higher capacity of freshwater to maintain metals in the water column, either in solution or in suspension.[43] Also, under hyposaline conditions, the concentration of free metal ions, e.g., Zn^{2+}, which are most readily taken up, would be at their maximum.[42] Second, different salinities may cause different rates of metal uptake due either to a linkage of ion fluxes across the body surface of an organism, or to physiological changes in the organism itself, such as drinking or water-filtration rates.[43]

Wright[44] has pointed out that observed salinity effects for cadmium may, in fact, be fundamentally due to the calcium concentration of the external medium. If trends in calcium levels parallel salinity, elevated cadmium accumulation could result from occupation of calcium binding sites by cadmium, under low calcium conditions.

The situation with regard to bioaccumulation under rapidly fluctuating salinities is more complex. Such conditions can elicit valve closure in marine bivalves, affecting uptake behavior and producing erratic bioaccumulation factors. The bioaccumulation of metals from food sources can also be influenced by salinity, as metal adsorption to phytoplankton is affected by varying salinity.

C. ORGANIC MATTER

Naturally occurring organic ligands, such as humic and fulvic acids, can form complexes and chelates with metals. This complexation increases the apparent solubility of metals, and also influences the extent to which they may be absorbed onto particulate matter. Thus, organic matter plays an important role in the transport and availability of metals in aquatic systems.[45] Metal availability for bioaccumulation is also dependent upon the size of the organic fraction with which it is associated, the distribution of which can vary seasonally.[46]

Laegreid et al.[47] found that low molecular weight organics may increase metal uptake by algae, while a similar result has been observed for zooplankton with low molecular weight humic acids. The effect of size was examined more closely by Giesy et al.,[46] who separated naturally occurring organics in pond water into four nominal diameter size fractions (I, > 0.0183 µm; II, 0.0183 to 0.0032 µm; III, 0.0032 to 0.0009 µm, IV < 0.0009 µm) by membrane ultrafiltration. ^{241}Am uptake by algae and bacteria were reduced by fractions II, III, and IV, although a growth dilution effect was suggested.[18]

Comparatively little attention has been given to analogous studies with higher organisms, such as fish. The smallest size organic fraction was found to increase cadmium toxicity towards *Gamburia affinis*, though this does not necessarily imply increased bioaccumulation of cadmium.[46] Clearly, further work is required to fully understand the effect of organic matter on metal bioaccumulation.

D. pH

As found earlier in this chapter, the pH has a major influence on the speciation of metals in water. Any discussion of the effect of pH on bioaccumulation is, in reality, largely one of the bioaccumulation potential of the various metal species. With algae, the influence of pH is reported to be species and metal specific. Cadmium levels in *Chlorella pyrenoidosa* cultures grown at a pH of 7 are greater than those obtained at a pH of 8. *Chlamydomonas* sp. bioaccumulate lead most efficiently at pH 4 to 9, whereas *Ulothrix fimbrinata* bioaccumulated lead to a significant degree only at a pH of 9. This differential behavior is attributable to variation in the composition of the cell surfaces of algae.[48]

Data for fish suggest that a pH influence may be metal specific also. Increased lead levels of rainbow trout are observed in acidic water due to increased branchial permeability. With copper, however, uptake is reduced at low pH values, due to increased mucus production of the gills.[42]

E. CHELATORS AND SURFACTANTS

The chemical forms in which metals are present influence their bioaccumulation. Since chelators and some surfactants can complex metals, it might be expected that their presence would affect metal bioaccumulation.

With the marine diatom, *Thalassiosira rotula*, the amount of cadmium and nickel taken up was not influenced by the presence of 1 × 10^5 M ethylenediaminetetraacetic acid (EDTA).[49] Studies with *Daphnia magna* have shown, however, that the presence of chelators such as $Na_4P_2O_7$, nitriloacetic acid (NTA), EDTA and EBDP (a structural analogue of EDTA), was effective in reducing biotic cadmium levels. Alternatively, cadmium in the presence of diethyldithiocarbamate (DCC) bioaccumulated to a greater degree than in its absence. This behavior may reflect the relatively greater hydrophobic character of the cadmium-DCC complex.[50]

Complexation of cadmium with either EDTA, pectin, or alginate (the major polysaccharide of brown seaweeds) doubled the final tissue concentrations in the mussel *Mytilus edulis*, and eliminated lag times for bioaccumulation. This suggests that a ligand-mediated uptake mechanism is operating.[15]

Surfactants are known to alter the permeability of biological membranes, and in this way could modify the bioaccumulation of metals. Gill perfusion experiments with rainbow trout have shown that anionic surfactants increased cadmium uptake at sublethal metal concentrations. In contrast, no effect is observed with non-ionic surfactants, indicating a specific mode of action for anionic surfactants, such as interaction with membrane-bound proteins involved in cadmium transport.[51]

F. OTHER METALS

In the natural environment metals are rarely found in isolation. Rather, they are usually encountered in association with many other metals. Any discussion of the bioaccumulation of metals would be incomplete without consideration of this influence. Unfortunately, our knowledge of this topic is fragmentary, and no overall trends can be clearly discerned.

With algae, the presence of copper, cadmium, and manganese has been observed to reduce the bioaccumulation of ^{65}Zn by *Laminaria digitata*. Cadmium bioaccumulation in a *Chlorella pyrenoidosa* culture was found to be unaffected by cobalt, copper, molybdenum, and zinc, but almost completely inhibited by manganese. This behavior may be due to the possession of semiselective binding sites which are capable of occupancy by a particular metal, depending upon its concentration relative to other metals also capable of occupancy.[48,52]

Similar competitive behavior is seen with higher organisms. The addition of $ZnCl_2$ to seawater markedly reduced cadmium tissue levels in bivalves with respect to controls containing no zinc.[39] High levels of calcium in ambient water have been demonstrated to decrease cadmium bioaccumulation in both freshwater and marine crustaceans. Since there is no known physiological function of cadmium, a competition with calcium in a calcium uptake or regulatory pathway is suggested.[17,44]

Calcium ion concentrations, together with magnesium ion concentrations, are critical determinants of the hardness of water. Based on the above experiments with crustaceans, it might be expected that bioaccumulation factors would be greatest in soft water. This expectation is borne out in comparative uptake studies of cadmium with snails, catfish, and guppies in hard (total Ca^{2+} and Mg^{2+} approximately 150 $\mu g\ g^{-1}$) and soft (total Ca^{2+} and Mg^{2+} O $\mu g\ g^{-1}$) water.[53] Thus, various measures of water quality, such as hardness, are important variables influencing the bioaccumulation behavior of metals.

V. EXAMPLES OF BIOACCUMULATION

A. CADMIUM

There is an extensive volume of literature relating to the bioaccumulation of metals. It is not intended to present an exhaustive compilation of this information in this section, but rather selected examples, so that similarities and differences between diverse groups of organisms can be identified.

Cadmium is primarily found associated with zinc, which is not surprising since they occur in the same group of the Periodic Table. It is produced commercially as a by-product

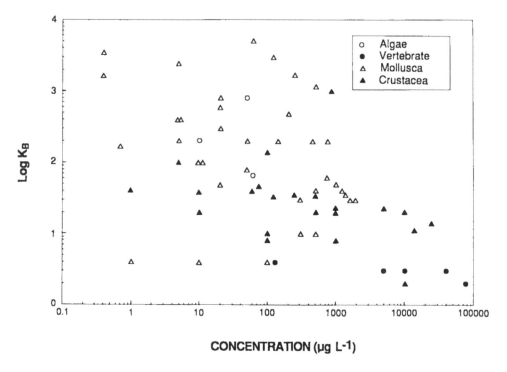

FIGURE 3. A plot of the logarithm of the bioconcentration factor for cadmium in freshwater as a function of aqueous concentration, with selected freshwater organisms. (From Taylor, D., *Ecotoxicol. Environ. Saf.*, 7, 43, 1983. Copyright Academic Press. With permission.)

of zinc smelting for use as a stabilizer and pigment in plastics, in electroplating, in various alloys, and in batteries.[2]

On the basis of more than forty laboratory investigations, cadmium bioconcentration factors for both marine and freshwater algae, vertebrates, molluscs, and crustaceans have been compiled.[26] The logarithms of these are plotted against aqueous concentration, with the freshwater data presented in Figure 3 and the marine in Figure 4. The bioconcentration factors relate either to steady state equilibrium conditions, or where such conditions were not reached, to the maximum value attained at the end of the exposure period.

The data in Figure 3 for freshwater organisms reveal that algae consistently exhibit the highest bioconcentration factors. Crustaceans, on the other hand, show log K_B values ranging between 1 and 3, while vertebrate bioconcentration factors, although variable, are usually the lowest. In fact, some 60% of the vertebrate (fish) K_B values reported were less than 20. This is consistent with other experimental results showing that freshwater invertebrates, such as snails and amphipods, accumulate up to three times more cadmium compared with vertebrate species.[54] The calculated median bioconcentration factor for the test organisms in Figure 3 is 90.

There are comparatively more data available concerning cadmium bioconcentration by marine biota.[26] From Figure 4, molluscs tend to bioaccumulate the most cadmium. This is species dependent, as illustrated by Grieg,[55] who found that oysters accumulated cadmium, copper, and silver to greater levels than both surf clams and ocean quahogs.[55] Crustaceans exhibit the next highest log K_B values, with almost all being less than 2. Furthermore, little variation in K_B with seawater cadmium concentration is observed. As in freshwater, marine vertebrates do not bioconcentrate cadmium to any large extent.

Some regulatory authorities such as the Japanese Ministry of International Trade and Industry (MITI) consider that chemicals with bioconcentration factors greater than 1,000

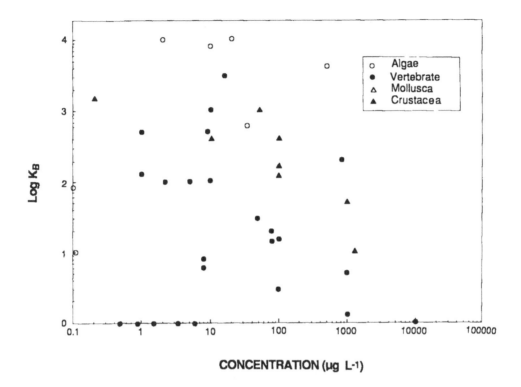

FIGURE 4. A plot of the logarithm of the bioconcentration factor for cadmium in seawater as a function of aqueous concentration, with selected marine organisms. (From Taylor, D., *Ecotoxicol. Environ. Saf.*, 7, 43, 1983. Copyright Academic Press. With permission.)

may represent an environmental hazard. Conversely, chemicals possessing bioconcentration factors less than 100 are considered to represent little threat.[26] Using these criteria, of the organisms considered in Figures 3 and 4, a potential environmental hazard may exist with cadmium in marine molluscs. To date, there have been a number of examples of poisoning attributed to the consumption of contaminated molluscs.[2]

As a caveat, it should be remembered that any bioconcentration factors employed for comparative purposes should be true equilibrium values. This is particularly important in any interspecies comparison of observed bioconcentration factors, since equilibrium times may vary considerably. *Daphnia magna*, for instance, attain equilibrium in static systems within 3 to 4 d, compared with 7 d for catfish and guppies.[33,50] Under flow-through conditions, bass and bluegill have been reported to attain equilibrium after 2 months exposure.[16]

Caution should also be exercised when comparing bioaccumulation factors of cadmium or any other metal, since primary uptake pathways may differ. Not only will overall tissue concentrations vary according to source, but the internal distribution of accumulated metal will also change. Studies on the uptake of cadmium by the crab *Carcinus maenus* have shown that the exoskeleton contains 80% of the total cadmium body burden when bioconcentrated from solution, but only 22% when food is the sole source.[56]

B. MERCURY

Mercury is most often encountered in nature as mercuric sulphide or cinnibar. A number of natural processes release a continuous but relatively minor amount of mercury into the aquatic environment.[9] Man has used mercury compounds for a multitude of purposes, such as production of chlorine and caustic soda, and as fungicides. Fossil fuels contain low levels of mercury, but since they are burnt on such a large scale, make a considerable contribution to atmospheric, and eventually, aquatic levels. Following the discovery in the early 1960s of the potential dangers to human health associated with mercury in aquatic areas, a reduction in input from major anthropogenic sources has occurred.[2]

Algal bioconcentration of mercury appears to be species specific. Bioconcentration factors for a diverse group of algae[57] have been shown to range from negligible to greater than 4×10^4.

As with other metals, molluscs can bioconcentrate mercury very efficiently. Highest concentrations are usually found in the viscera. The bivalve *Crassostrea virginica*, when exposed to 100 ng g^{-1} mercuric acetate, achieved a tissue concentration of 1×10^5 ng g^{-1} after 45 d, a bioconcentration factor of 10^3. In depuration experiments, the rate of loss often depends on the pathway by which the mercury was acquired. It is lost rapidly if bioconcentrated, and more so if food was the source; however, mercuric chloride injected directly into the foot muscle of molluscs is lost only very slowly.[2]

Mercury may enter fish via the gills, or from dietary sources. Studies of methyl mercuric chloride bioaccumulation with rainbow trout have suggested that the amount of mercury accumulated from both sources is quantitatively additive.[8] Relatively few mercury bioaccumulation or bioconcentration factors for fish are found, because the biotic concentration increases with size in a well-defined relationship within a species. Such concentrations should be normalized with respect to size.[58] A total mercury bioconcentration factor range with fish of 10^4 to 10^5 has been suggested.[59] Some large oceanic fish, such as halibut, swordfish, and marlin, contain relatively high concentrations of mercury; these fish are large carnivores at the end of a food chain, and they are very active fish, swimming with their mouths open, producing a forced flow of water containing dissolved mercury across the gills.[2] Additionally, much of the mercury contained in fish is in the form of MeHgX, which has a very long half life, of the order of years.[9]

Marine mammals such as seals, sea lions, and dolphins bioaccumulate large amounts of mercury without apparent toxic effect. Interestingly, trends in tissue mercury levels seem to be paralleled by selenium levels, and the mercuric selenide found in the liver of these animals may be the product of a detoxifying mechanism.

C. COPPER

Copper is an essential element for animals; the highest concentrations are usually found in decapod crustaceans, gastropods, and cephalopods in which the respiratory pigment haemocyanin contains copper.[2] Copper concentrations in algae from costal waters have been found to reflect ambient water concentrations, and some species (e.g., *Fucus vesiculosus* and *Ascophyllum nodosum*) are suggested as potential biological indicators for the metal. Bioconcentration factors for brown algae[52] average approximately 1×10^4.

Molluscs, and in particular oysters, have a remarkable facility to accumulate copper. Laboratory experiments show that bioconcentration factors of 10^2 to 10^3 for clams and mussels, and 10^4 to 10^5 for oysters are not unusual.[44,45] Limpets store copper as metalloprotein complexes in the liver and kidney, and this may serve to explain the general tolerance of molluscs in copper-contaminated environments, and the extremely high levels accumulated under such conditions.[32] Copper taken up by mussels (*Mytilus edulis*) is also observed to cause a displacement of aluminum and molybdenum from all organs.[45] On the basis of a relationship between oyster size and copper content, Ayling[20] has contended that copper is bioaccumulated by these organisms via a physiological process governed by the size of the oyster.

Algae, molluscs, crustaceans, and fish all bioaccumulate copper, but the metal does not seem to biomagnify through food chains. A predatory fish such as the marlin, at the top of a chain chain, accumulates mercury, but not copper.[2]

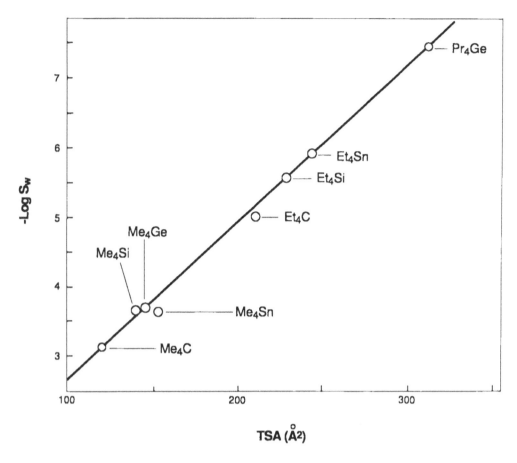

FIGURE 5.   The logarithm of the maximum aqueous solubility of tetraalkyl derivatives of Group IVA as a function of TSA. (From Craig, P. J., *Organometallic Compounds in the Environment: Principles and Reactions*, 1986, Chap. 2. Copyright John Wiley & Sons. With permission.)

VI. BIOACCUMULATION OF ORGANOMETALLIC COMPOUNDS

There is an increasing use of organometallic chemicals, such as organomercury and organotin compounds. Organomercurials are used as fungicides and slimicides, while organotin compounds find considerable use as polymer stabilizers, catalytic agents, and biocides in coating processes and in antifouling paints. A recent report suggests that approximately 10% of the total organotin compounds used in antifouling paints and as biocides enter the aquatic environment, including sediment.[60]

Within this environment, there are additional biogenic sources of organometallics. Both lead and mercury are bio-alkylated, probably by microorganisms, while the organoarsenic derivatives found in large amounts in many shellfish are synthesized primarily by phytoplankton, and then passed along the food chain without significant modification.[34]

Despite the prevalence of organometallic species, relatively little is known about their environmental behavior in general, and bioaccumulation behavior in particular. It is logical, however, that an increased number of organic substituents would tend to confer a greater hydrophobicity on an organometallic compound, leading to a greater tendency for bioaccumulation. In support of this the uptake behavior of organomercurials (R_2Hg) has been considered to resemble that of organochlorines. The importance of the increased lipid solubility and decreased aqueous solubility of these compounds has also been shown by the correlation between their toxicity towards crustaceans and diethyl ether/seawater partition coefficients.[34]

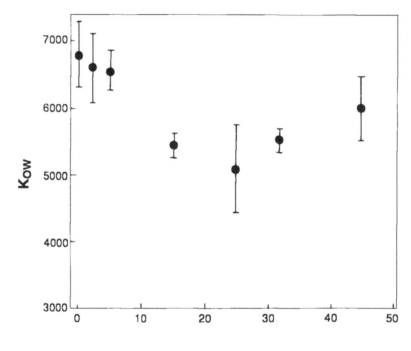

SALINITY (parts per thousand)

FIGURE 6. The octanol/water partition coefficient (K_{OW}) of tributyltin species as a function of salinity. The bars represent ± 1 standard deviation. (From Laughlin, R. B., Guard, H. E., and Coleman, W. M., *Environ. Sci. Technol.*, 20, 201, 1986. Copyright American Chemical Society. With permission.)

Not only the number, but also the nature of the organic substituent would be expected to influence bioaccumulation behavior of organometallics. In view of the diversity of these substituent groups, the prediction of bioaccumulation potential represents a major problem. Total surface areas (TSA) of related organometallic compounds have been shown to be reliable estimators of toxicity.[59,61] Aqueous solubilities have also been correlated with calculated TSA, as shown in Figure 5. Since aqueous solubility, and hence TSA, is linearly related to bioconcentration factor with nondegradable hydrophobic organic contaminants, it follows that a similar relationship may be applicable here.

The use of K_{OW} to predict K_B values for organometallics is complicated by the speciation processes of these compounds. For example, on dissolution in seawater the common antifouling agent $[Bu_3Sn]_2O$ forms an equilibrium mixture comprising Bu_3SnCl, Bu_3SnOH, the aqua complex $[Bu_3SnOH_2]^+$, and a tributyltin carbonate species. This distribution is influenced by chloride ion and dissolved CO_2 concentration, as well as pH, and is easily displaced by variation within the normal environmental range of these parameters. Thus, it is more appropriate to describe the K_{OW} value as being for the $[Bu_3Sn]^+$ species, and the conditions should always be accurately specified. The variation of K_{OW} with salinity for $[Bu_3Sn]^+$ is depicted in Figure 6. It shows K_{OW} to be relatively invariant between 15 and 30 parts per thousand salinity, but with an increasing K_{OW} value as the salinity approaches zero.[62] This suggests that higher bioconcentration factors may be obtained with tributyltin compounds in freshwater compared to seawater.

As an example of the bioaccumulation behavior of organometallic compounds, K_B values for carp (*Cyprinus carpio*) with Ph_3SnCl, Ph_2SnCl_2 and $PhSnCl_3$ have been determined under flowthrough conditions. Measured log K_{OW} values for these species are 2.11, 1.43, and

1.73, and bioconcentration equilibrium times paralleled this order. Results showed that in muscle, liver, kidney, and gallbladder tissues bioconcentration factors also followed the log K_{OW} order.[63] Some biodegradation of Ph_3SnCl was noted, and it appears that organotin compounds, at least, are degraded by cytochrome P-450 dependent mono-oxygenase enzyme systems in a manner similar to that observed with alkanes and polyaromatic hydrocarbons.[64]

A comparative study of bioaccumulation pathways involving the mussel *Mytilus edulis* has revealed that tributyltin compounds can either be bioconcentrated or bioaccumulated from a food source, such as phytoplankton. Tributyltin species accumulated differentially in gills, mantle muscle, and viscera, but regardless of source, tissue burdens correlated well with lipid content, supporting a partitioning mechanism for bioaccumulation.[65] On a whole body basis, following 30 d exposure to contaminated phytoplankton, the observed bioaccumulation factor was less than 2. Following 47 d bioconcentration, the observed K_B value was approximately 5×10^3. Similar values have been obtained for tributyltin derivatives with other molluscs. On the basis of published relationships between log K_B and log K_{OW} for persistent organic compounds with molluscs, a bioconcentration factor of 1×10^2 would be anticipated.[66] No theory proposed to date has been adequate to explain this apparent anomaly in bioconcentration factors. Therefore, although organometallic compounds resemble nonpolar organic compounds in many respects, important differences exist because of unique physicochemical properties and chemical speciation, which are reflected in bioaccumulation behavior.[65]

REFERENCES

1. **Engel, D. W. and Fowler, B. A.,** Factors influencing cadmium accumulation and its toxicity to marine organisms, *Environ. Health Perspect.*, 28, 81, 1979.
2. **Clark, R. B.,** *Marine Pollution*, Claredon Press, Oxford, 1986, Chap. 6.
3. **Williams, S. L., Aulenbach, D. B., and Clesceri, N. L.,** Sources and distribution of trace metals in aquatic environments, in *Aqueous Environmental Chemistry of Metals*, Rubin, A, J., Ed., Ann Arbor Science Publishers, Ann Arbor, MI, 1976, Chap. 2.
4. **Skinner, B. J. and Terekian, K. K.,** *Man and the Ocean*, Prentice-Hall, Englewood Cliffs, New Jersey, 1973, Chap. 2.
5. **Bryan, G. W.,** The effects of heavy metals (other than mercury) on marine and estuarine organisms, *Proc. R. Soc. London, Ser. B.*, 177, 389, 1971.
6. **Bryan, G. W.,** Heavy metal contamination in the sea, in *Marine Pollution*, Johnston, R., Ed., Academic Press, London, 1976, Chap. 3.
7. **Wright, D. A.,** Heavy metal accumulation by aquatic invertebrates, *Appl. Biol.*, 3, 331, 1978.
8. **Phillips, G. R. and Buhler, D. R.,** The relative contributions of methylmercury from food or water to rainbow trout (*Salmo gairdneri*) in a controlled laboratory environment, *Trans. Am. Fish. Soc.*, 107, 853, 1978.
9. **Peterson, C. L., Klawe, W. L., and Sharp, G. D.,** Mercury in tunas: A review, *Fish. Bull.*, 71, 603, 1973.
10. **Zirino, A. and Yamamato, S.,** A pH-dependent model for the chemical speciation of copper, zinc, cadmium and lead in seawater, *Limnol. Oceanogr.*, 17, 661, 1972.
11. **Chakoumakos, C., Russo, R. C., and Thurston, R. V.,** Toxicity of copper to cutthroat trout (*Salmo clarki*) under different conditions of alkalinity, pH and hardness, *Environ. Sci. Technol.*, 13, 213, 1979.
12. **Canterford, G. S., Buchanan, A. S., and Ducker, S. C.,** Accumulation of heavy metals by the marine diatom *Ditylum brightwelli* (West) Grunow, *Aust. J. Mar. Freshwater Res.*, 29, 613, 1978.
13. **Wang, H-K. and Wood, J. M.,** Bioaccumulation of nickel by algae, *Environ. Sci. Technol.*, 18, 106, 1984.
14. **Brooks, R. R. and Rumsby, M. G.,** The biogeochemistry of trace element uptake by some New Zealand bivalves, *Limnol. Oceanogr.*, 10, 521, 1965.
15. **George, S. S. and Coombs, T. L.,** The effects of chelating agents on the uptake and accumulation of cadmium by *Mytilus edulis*, *Mar. Biol.*, 39, 261, 1977.

16. **Cearley, J. E. and Coleman, R. L.**, Cadmium toxicity and bioconcentration in largemouth bass and bluegill, *Bull. Environ. Contam. Toxicol.*, 11, 146, 1974.

17. **Wright, D. A.**, Cadmium and calcium interactions in the freshwater amphipod *Gammarus pulex*, *Freshwater Biol.*, 10, 123, 1980.

18. **Giesy, J. P. and Paine, D.**, Effects of naturally occurring aquatic organic fractions on [241]Am uptake by *Scenedesmus obliquus (Chlorophyceae)* and *Zeromonas hydrophila (Pseudomonadaceae)*, *Appl. Environ. Microbiol.*, 33, 89, 1977.

19. **Pentreath, R. J.**, The accumulation from water of [65]Zn, [54]Mn, [58]Co and ^{59}Fe by the mussel, *Mytilus edulis*, *J. Mar. Biol. Assoc.*, (UK), 53, 127, 1973.

20. **Ayling, G. M.**, Uptake of cadmium, zinc, copper, lead, and chromium in the Pacific oyster, *Crassostrea gigas*, grown in the Tamar River, Tasmania, *Water Res.*, 8, 729, 1974.

21. **Bryan, G. W.**, Bioaccumulation of marine pollutants, *Phil. Trans. R. Soc. London, Ser. B.*, 286, 483, 1979.

22. **Memmert, U.**, Bioaccumulation of zinc in two freshwater organisms (*Daphnia magna*, Crustacea and *Brachydanio rerio*, Pisces), *Water Res.*, 21, 99, 1987.

23. **St. John, B. E.**, Heavy metals in the skeletal carbonate of scleractinian corals, *Proc. 2nd. Int. Coral Reef Symp.*, Great Barrier Reef Committee, Brisbane, 2, 461, 1974.

24. **Hatakeyama, S. and Yasuno, M.**, Accumulation and effects of cadmium on guppy (*Poecilia reticulata*) fed cadmium-dosed Cladocera (*Moina macrocopa*), *Bull. Environ. Contam. Toxicol.*, 29, 159, 1982.

25. **Ferard, J. F., Jouany, J. M., Trukaut, R., and Vasseur, P.**, Accumulation of cadmium in a freshwater food chain experimental model, *Ecotoxicol. Environ. Saf.*, 7, 43, 1983.

26. **Taylor, D.**, The significance of the accumulation of cadmium by aquatic organisms, *Ecotoxicol. Environ. Saf.*, 7, 33, 1983.

27. **Nimmo, D. W. R., Lightner, D. V., and Bahner, L. H.**, Effects of cadmium on the shrimps *Penaeus duorarum*, *Palaemontes pugio* and *Palaemontes vulgaris*, in *Physiological Responses of Marine Biota to Pollution*, Vernberg, J. F., Ed., Academic Press, New York, 1977, 131.

28. **Tessier, A., Campbell, P. G. C., Auclair, J. C., and Bisson, M.**, Relationships between the partitioning of trace metals in sediments and their accumulation in the tissues of the freshwater mollusc *Elliptio complanata* in a mining area, *Can. J. Fish. Aquatic Sci.*, 41, 1463, 1984.

29. **Miramand, P., Germain, P., and Camus, H.**, Uptake of Americium and Plutonium from contaminated sediments by three benthic species: *Arenicola marina*, *Corophium volutator* and *Scrobicularia plana*, *Mar. Ecol. Prog. Ser.*, 7, 59, 1982.

30. **Klumpp, D. W. and Burdon-Jones, C.**, Investigations of the potential of bivalve molluscs as indicators of heavy metal levels in tropical marine waters, *Aust. J. Mar. Freshwater Res.*, 33, 285, 1982.

31. **Lake, P. S.**, Accumulation of cadmium in aquatic animals, *Chem. Aust.*, 46, 26, 1979.

32. **Howard, A. G. and Nickless, G.**, Heavy metal complexation in polluted molluscs 1. Limpets (*Patella vulgata* and *Patella intermedia*), *Chem. Biol. Interactions*, 16, 107, 1977.

33. **Williams, D. R. and Giesy, J. P.**, Relative importance of food and water sources to cadmium uptake by *Gambusia affinis*, *Environ. Res.*, 16, 326, 1978.

34. **Phillips, D. J. H.**, *Quantitative Aquatic Biological Indicators*, Applied Science Publishers, London, 1980, 85.

35. **Fagerstrom, T., Asell, B., and Jernelov, A.**, Model for accumulation of methyl mercury in northern pike (*Esox lucuis*), *Oikos*, 25, 14, 1974.

36. **Majori, L. and Petronio, F.**, Marine pollution by metals and their accumulation by biological indicators (Accumulation factor), *Rev. Intern. Oceanogr. Med.*, 31—32, 55, 1973.

37. **Ruzic, I.**, Two-compartment model of radionuclide accumulation into marine organisms. I. Accumulation from a medium of constant activity, *Mar. Biol.*, 15, 105, 1972.

38. **Popham, J. D. and D'Auria, J. M.**, Statistical models for estimating seawater metal concentrations from metal concentrations in mussels (*Mytilus edulis*), *Bull. Environ. Contam. Toxicol.*, 27, 660, 1981.

39. **Jackim, E., Morrison, G., and Steele, R.**, Effects of environmental factors on radiocadmium uptake by four species of marine bivalves, *Mar. Biol.*, 40, 303, 1977.

40. **Rodgers, D. W. and Beamish, F. W. H.**, Uptake of water borne methylmercury by rainbow trout (*Salmo gairdneri*) in relation to oxygen consumption and methylmercury concentration, *Can. J. Fish. Aquatic Sci.*, 38, 1309, 1981.

41. **Phillips, D. J. H.**, The common mussel *Mytilus edulis* as an indicator of pollution by zinc, cadmium, lead and copper. 1. Effects of environmental variables on uptake of metals, *Mar. Biol.*, 38, 59, 1976.

42. **Part, P., Svanberg, O., and Kiessling, A.**, The availability of cadmium to perfused rainbow trout gills in different water qualities, *Water Res.*, 19, 427, 1985.

43. **Phillips, D. J. H.**, Effects of salinity on the net uptake of zinc by the common mussel *Mytilus edulis*, *Mar. Biol.*, 41, 79, 1977.

44. **Wright, D. A.**, The effect of calcium on cadmium uptake by the shore crab *Carcinus maenas*, *J. Exp. Biol.*, 67, 163, 1977.

45. **Sutherland, J. and Major, C. W.**, Internal heavy metal changes as a consequence of exposure of *Mytilus edulis*, the blue mussel, to elevated external copper (II) levels, *Comp. Biochem. Physiol.*, 68C, 63, 1981.

46. **Giesy, J. P., Leversee, G. J., and Williams, D. R.**, Effects of naturally occurring aquatic organic fractions on cadmium toxicity to *Simocephalus serrulatus* (Daphnidae) and *Gambusia affinis* (Poeciliidae), *Water Res.*, 11, 1013, 1977.

47. **Laegreid, M., Alstad, J., Klaveness, D., and Seip, H. M.**, Seasonal variation of cadmium toxicity toward the algae *Selenastrum capricornutum* Printz in two lakes with different humus content, *Environ. Sci. Technol.*, 17, 357, 1983.

48. **Hart, B. A. and Scaife, B. D.**, Toxicity and bioaccumulation of cadmium in *Chlorella pyrenoidosa*, *Environ. Res.*, 14, 401, 1977.

49. **Dongmann, G. and Nurnberg, H. W.**, Observations with *Thalassiosira rotula* (Meunier) on the toxicity and accumulation of cadmium and nickel, *Ecotoxicol. Environ. Saf.*, 6, 535, 1982.

50. **Poldoski, J. E.**, Cadmium bioaccumulation assays. Their relationship to various ionic equilibria in Lake Superior water, *Environ. Sci. Technol.*, 13, 701, 1979.

51. **Part, P., Svanberg, O., and Bergstrom, E.**, The influence of surfactants on gill physiology and cadmium uptake in perfused rainbow trout gills, *Ecotoxicol. Environ. Saf.*, 9, 135, 1985.

52. **Foster, P.**, Concentrations and concentration factors of heavy metals in brown algae, *Environ. Pollut.*, 10, 45, 1976.

53. **Kinkade, M. L. and Erdman, H. E.**, The influence of hardness components (Ca^{2+} and Mg^{2+}) in water on the uptake and concentration of cadmium in a simulated freshwater ecosystem, *Environ. Res.*, 10, 308, 1975.

54. **Spehar, R. L., Anderson, R. L., and Fiandt, J. T.**, Toxicity and bioaccumulation of cadmium and lead in aquatic invertebrates, *Environ. Pollut.*, 15, 195, 1978.

55. **Greig, R. A.**, Trace metal uptake by three species of mollusks, *Bull. Environ. Contam. Toxicol.*, 22, 643, 1979.

56. **Jennings, J. R. and Rainbow, P. S.**, Studies on the uptake of cadmium by the crab *Carcinus maenus* in the laboratory. I. Accumulation from seawater and food source, *Mar. Biol.*, 50, 131, 1979.

57. **Hasset, J. M., Jennet, J. C., and Smith, J. E.**, Microplate technique for determining accumulation of metals by algae, *Appl. Environ. Microbiol.*, 41, 1097, 1981.

58. **Barber, R. T., Whaling, P. J., and Cohen, D. M.**, Mercury in recent and century-old deep-sea fish, *Environ. Sci. Technol.*, 18, 552, 1984.

59. **Craig, P. J.**, *Organometallic Compounds in the Environment. Principles and Reactions*, John Wiley & Sons, New York, 1986, Chap. 2.

60. **Donard, O. F. X. and Weber, J. H.**, Behaviour of methyltin compounds under simulated estuarine conditions, *Environ. Sci. Technol.*, 19, 1104, 1985.

61. **Laughlin, R. B., Johannesen, R. B., French, W., Guard, H., and Brinckman, F. E.**, Structure-activity relationships for organotin compounds, *Environ. Toxicol. Chem.*, 4, 343, 1985.

62. **Laughlin, R. B., Guard, H. E., and Coleman, W. M.**, Tributyltin in seawater: Speciation and octanol-water partition coefficient, *Environ. Sci. Technol.*, 20, 201, 1986.

63. **Tsuda, T., Nakanishi, H., Aoki, S., and Takebayashi, J.**, Bioconcentration and metabolism of phenyltin chlorides in carp, *Water Res.*, 21, 949, 1987.

64. **Fish, F. H., Kimmel, E. C., and Casida, J. E.**, Bio-organotin chemistry: Reactions of tributyltin derivatives with a cytochrome P-450 dependent monooxygenase enzyme system, *J. Organomet. Chem.*, 118, 41, 1976.

65. **Laughlin, R. B., French, W., and Guard, H. E.**, Accumulation of bis(tributyltin) oxide by the marine mussel *Mytilus edulis*, *Environ. Sci. Technol.*, 20, 884, 1986.

66. **Hawker, D. W. and Connell, D. W.**, Bioconcentration of lipophilic compounds by some aquatic organisms, *Ecotoxicol. Environ. Saf.*, 11, 184, 1986.

Chapter 9

ACHIEVEMENTS AND CHALLENGES IN BIOACCUMULATION RESEARCH

Des W. Connell

TABLE OF CONTENTS

I. BIOACCUMULATION PATHWAYS AND MECHANISMS FOR LIPOPHILIC COMPOUNDS

The overall pathway for the bioaccumulation of xenobiotic lipophilic chemicals in the environment can be summarized as shown in Figure 1. This set of pathways indicates the fundamental importance of sediments in aquatic areas and the soil in terrestrial areas, as a reservoir of xenobiotic chemicals which supplies phases such as ambient water, interstitial water, water in the gastrointestinal tract, and air. These latter phases usually contain low concentrations of chemicals, but are dynamic in supplying organisms with chemicals through the lungs, gills, oxygen uptake surfaces, or the gastrointestinal tract.

In both terrestrial and aquatic animals, a somewhat similar set of processes operates. Uptake through lungs, gills, oxygen uptake surfaces, or the gastrointestinal tract is followed by the establishment of an equilibrium with the circulatory fluid, which in turn, equilibrates with body lipids. However, major differences between terrestrial and aquatic organisms are the clearance patterns. Aquatic organisms can excrete unchanged chemicals to a large water mass, whereas with terrestrial organisms the corresponding route through the lungs to the ambient air is very slow with many higher molecular xenobiotic compounds. As a result, terrestrial organisms tend to favor clearance by metabolism to hydrophilic compounds and excretion. In addition, uptake of these compounds through the lungs is not usually significant due to their low atmospheric concentration, vapor pressure, and other factors.

Most of these transfers are governed by equilibrium processes between different phases. In many cases, the equilibria are between similar lipoid material in different phases, for example, between circulatory fluid and body lipids. In other cases, the equilibrium is between an aqueous and a lipoid phase, such as that between ambient water and aquatic organism lipid. These equilibria can provide the basis for quantitative investigation of bioaccumulation.

In both aquatic and terrestrial vegetation, uptake and clearance seem to be substantially controlled by uptake on the outer surfaces followed by diffusion to the plant lipids. Equilibrium between the plant lipids and the external water or atmosphere can probably be established, given a sufficient time period. Direct transfer of compounds taken up on the roots of plants through to the upper parts of the plant does not seem to be a significant process.

A number of suggestions have been made in this book regarding the transfer pathways of compounds in food chains. Generally, these require further evaluation and production of more confirmatory quantitative evidence. The theoretical basis of these food chain transfer mechanisms, which actually occur in nature, also require development and validation.

A particular area requiring similar investigation is the nature of the equilibrium between the contents of the gastrointestinal tract and the circulatory fluid and body lipids. The rate of food consumption, the physical state of the food, and several other factors may have an influence on this process. Other areas requiring attention are the nature of the soil to the atmosphere and also the vegetation-to-atmosphere equilibrium. Only a preliminary understanding of these equilibria and the factors influencing them are available at present. Similarly, the importance of the uptake and clearance of xenobiotic chemicals through the lungs by animals needs to be evaluated.

The rate at which all of these equilibria and transfers occur can have a significant effect on the observed behavior of bioaccumulated chemicals. Generally, the data base on the kinetics of these processes is limited. In addition, an understanding of the factors governing the kinetics is generally not available.

II. QSARS FOR BIOACCUMULATION

Previously, the importance of equilibrium processes in governing the behavior of li-

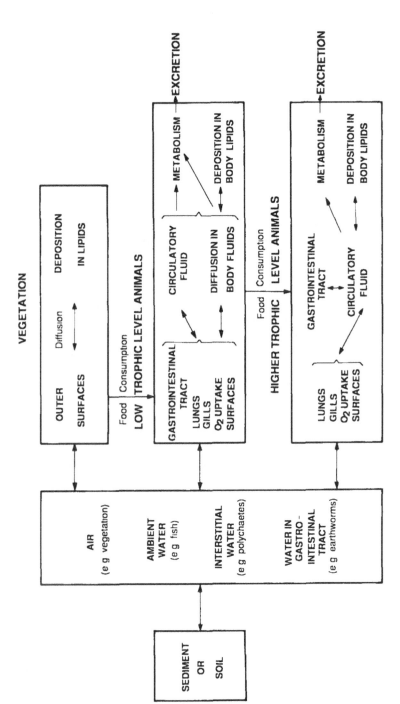

FIGURE 1. Diagrammatic illustration of the bioaccumulation pathways of xenobiotic chemicals in the environment.

pophilic chemicals in biota was outlined. The biotic lipids are the major biological components in which lipophilic compounds are deposited, and this is controlled substantially by partition with water phases. It is therefore not surprising that the octanol-to-water partition coefficient has been generally found to be the most useful predicator of bioaccumulation in many different types of biota. This is of particular value with a range of aquatic organisms. The available evidence suggests that the following general equations should be approximately applicable to most fully aquatic biota:

$$K_{BL} = K_{OW}$$

where K_{BL} is the bioconcentration factor expressed in terms of lipid weight,

$$\text{and} \qquad K_B = y_L \, K_{OW}$$

$$\text{or} \qquad \log K_B = \log K_{OW} + \log y_L$$

where K_B is the bioconcentration factor in terms of wet weight, and y_L is the proportion of lipid present in the organism. These relationships should apply irrespective of whether water or food is the source of the xenobiotic compound.

With biomagnification, the biomagnification factors are generally relatively low. In many cases these factors are related to the direct partitioning between lipoid material in food and body lipid contents. Often these contents of lipoid materials are similar, resulting in a low biomagnification factor. In addition, even though the biomagnification factors are usually low, a weak dependence on the octanol-to-water partition coefficient has been observed.

The previously discussed relationships are generally applicable to chlorinated hydrocarbons and polyaromatic hydrocarbons with aquatic organisms. Mammals and some other organisms exhibit higher biodegradation rates with many xenobiotic compounds, which introduces an additional factor depressing bioaccumulation capacity. Similarly, many compounds not in the two groups mentioned above exhibit significant biodegradation with fish. In these situations bioaccumulation capacity is reduced and cannot be predicted. There is a need for the development of methods to predict bioaccumulation of compounds which biodegrade utilizing physiochemical data, or data which can be readily obtained in the laboratory, or calculated.

There would be considerable application for quantitative structure-activity relationships (QSARs) of bioaccumulation based on parameters which can be conveniently calculated. These would allow bioaccumulation capacity to be estimated without conducting laboratory experiments. Considerable attention has been focused on fragmental methods for calculating the octanol-to-water partition coefficient. More recently, software has become available for calculation of a wide range of molecular descriptors, such as molecular surface area, Randic indices, and so on. These, either singly or in combination, offer considerable potential in this area which is, as yet, not fully developed.

Further validation is needed of the accuracy of present QSAR techniques for predicting bioaccumulation in actual environmental situations. A variety of factors may be important in field situations which are not taken into account in laboratory experiments, on which most bioaccumulation QSARs are based. For example, in the sediment-to-water process, in field situations colloids may exert a significant influence on observed relationships. Kinetic and exposure time periods and patterns may also exert an influence in field situations.

While QSARs for lipophilic compounds (log K_{OW} from 2 approximately 6) are reasonably well understood, those for superhydrophobic compounds (log K_{OW} > approximately 6) are not clear. There is a decline in the bioaccumulation capacity of superhydrophobic compounds, but how consistent this is for different organisms and different types of compound is not known. Two factors have been suggested to cause this decline, reducing fat solubility and membrane permeability, but clear evidence of the role of these factors, and possibly others, is not available.

III. BASIC THEORY OF PARTITIONING AND BIOCONCENTRATION

With partitioning between abiotic phases, a substantial body of theory has been developed. The basic importance of the molecular surface area and volume have been described. However, these factors alone do not provide an adequate basis for predicting the partition coefficient of a compound in different pairs of phases. Clearly, other factors are involved. Perhaps the inclusion of solute to solvent interactions and molecular stereochemistry may result in a model which will provide a more comprehensive description of the partitioning process. This more comprehensive model should be capable of predicting the partition coefficient of a compound from basic data on the compound itself and the two phases involved.

This theory of partitioning in a biotic phase should be used to develop a comprehensive theory of bioconcentration. This extended theory would need to take into account biological factors. Perhaps one of the most important of these is membrane permeation, although other factors would be needed as well. If the octanol-to-water partition coefficient were used in such a model, the differences between lipid and octanol would need to be taken into account. Similar to the abiotic partition theory mentioned above, such as extended theory should be capable of predicting the bioconcentration of a compound from basic data.

IV. BIOACCUMULATION OF METALLIC SUBSTANCES AND ORGANOMETALLIC COMPOUNDS

Generally, the available evidence indicates that organometallic compounds behave similarly to organic compounds on bioaccumulation. However, some important differences exist which require further evaluation. On the other hand, the bioaccumulation of other chemical forms of the metals, particularly the ionic forms, is less amenable to understanding. No presently available characteristic, comparable to the octanol-to-water partition coefficient, is available with which the bioaccumulation of these metallic forms can be predicted. This probably reflects the variety of mechanisms of deposition of metallic forms in the body tissues of organisms. Some may be deposited in bone tissue, others may be incorporated into particular organs, while others may have no presently known consistent pattern of deposition.

When more is known of the mechanism of bioaccumulation of metal forms, it may be possible to develop a predictive capacity. This may involve the use of physicochemical characteristics of the metal or compounds based on it. The physicochemical characteristics of most value would probably be those which bear a relationship to the deposition mechanism involved.

APPENDIX
COMMONLY USED SYMBOLS

a	Activity
A	Area
BF	Biomagnification factor
BHC	Hexachlorocyclohexane
C_A	Air concentration
C_B	Biotic concentration
C_{BM}	Maximum biotic concentration
C_C	Concentration in consumer
C_F	Concentration in food
C_l	Water solubility for liquids and super cooled solids
C_S	Sediment concentration
C_{SO}	Sediment concentration in organic carbon
C_V	Vegetation concentration
C_W	Water concentration
DDT	Commonly known as dichloro-diphenyl trichloroethane
D_M	Diffusion coefficient of solute in membrane
f	Fugacity
f	Fragmental constant for calculating K_{OW}
f_{oc}	Fraction of organic carbon
H	Henry's Law Constant
k'	Capacity factor
k_1	Uptake rate constant
k_2	Clearance role constant
K_B	Bioconcentration factor
K_{hepw}	Heptane to water partition coefficient
K_{hw}	Hexadecane to water partition coefficient
K_{MW}	Membrane to water partition coefficient
K_{LW}	Partition coefficient between lipid solvent and water
K_{OC}	Sediment to water partition coefficient in terms of organic carbon
K_{OM}	The soil to water partition coefficient in terms of organic matter
K_{OW}	Octanol to water partition coefficient
K_{SOA}	Soil to air partition coefficient
K_{SW}	Sediment to water partition coefficient also referred to as Kp and K_D

K_{VA}	Vegetation to atmosphere partition coefficient
l	Lipid fraction
L_V	Lipid fraction in vegetation
m	Dipole moment
M	Molecular weight
N_c	Number of carbon atoms
[P]	Parachor
P	Pressure
PAH	Polycyclic aromatic hydrocarbons
PCB	Polychlorobiphenyls
PCDD	Polychlorodibenzodioxin
PCDF	Polychlorodibenzofuran
ppb	Parts per billion
ppm	Parts per million
QSAR	Quantitative structure-activity relationship
R	Universal gas constant
S	Water solubility (mass volume^{-1})
S_M	Membrane thickness
S_o	Octanol solubility
S_W	Water solubility (moles volume^{-1})
t	Elapsed time period
t_{eq}	Time period to equilibrium
T	Temperature (°K)
TCDD	Tetrachlorodibenzodioxin
T_M	Melting point (°K)
TSA	Total molecular surface area
v	Environmental phase volume
V	Molar volume
V_M	Molar volume of water
x	Mole fraction
y	Compactness factor
y_L	Lipid fraction
z	Fugacity capacity constant
α	Polarizability
γ	Activity coefficient
φ	Volume fraction
μ	Chemical potential
χ	Florey-Huggins interaction parameter
O_x	Zero order connectivity index
1_x	1st Order connectivity index
Y*	Mole fraction activity coefficient
ρ	Density
σ	Molecular diameter

INDEX

Printed and bound by CPI Group (UK) Ltd, Croydon, CR0 4YY

22/10/2024

01777633-0012